EXERCISES IN BUILDING CONSTRUCTION

FOURTH EDITION

Forty-Five Homework and Laboratory
Assignments to Accompany

FUNDAMENTALS OF BUILDING CONSTRUCTION
MATERIALS AND METHODS
FOURTH EDITION

Edward Allen
and
Joseph Iano

WILEY

John Wiley & Sons, Inc.

The authors extend special thanks to Gale Beth Goldberg and Westley Spruill, who collaborated on earlier versions of this work.

Contents

The exercises in this book are designed to help you learn about materials and methods of construction by involving you in the kinds of work that building design professionals--architects, engineers, and drafters--do every day in the course of designing buildings and getting them built. You will find that these exercises make it easier to learn the essential information in the accompanying text, Fundamentals of Building Construction. You will also discover that they will give you a good start toward becoming proficient in many different phases of building activity.

Keep Fundamentals of Building Construction close by as you do the work in this book, and get in the habit of consulting it frequently. Nearly everything you need to know to solve the problems is in this textbook, and in most cases you will be given explicit directions about where to look for it. The glossary at the end of the text will be useful if you do not understand a technical term, and the index can help you locate information that is not directly referenced from the exercises.

These exercises are intended to be hand drafted. Despite the dominance of computer aided technology in production drawing, the skills you will develop in these exercises--to conceptualize and develop building assemblies with confidence and ease--remain fundamental. You may draw freehand or with the aid of a drafting board and instruments, as you prefer or as directed by your instructor. In either case, only minimal use of an architect's scale is required. You can scale your drawing using the squares of the printed grid and the scale designation at the lower corner of the page. Always complete your solution to the given scale. Only on pages where no scale is given should you not work to scale.

You will often be asked to draw a section detail of a building assembly such as a wall, column, floor, or beam. You will find that the easiest way to do this is to **draw the components of your detail in the**

same sequence in which they will be assembled in the actual building. First draw the basic structural components, then the major parts added to the structure, then the finish and trim pieces. This will help you learn the underlying logic of the detail, and thereby remember the detail more easily. Trying to learn a complex detail by staring at it and attempting to memorize its shapes is virtually impossible for most people, and is not at all useful in increasing your understanding or professional skills.

Block out each drawing on the page with light lines before you begin to draw final lines. Outline lightly all major components of your solution. If you are in doubt about what to do next, use tracing paper or scratch paper to test alternatives before you commit lines to the sheet you will turn in. When you are satisfied that you have everything right, darken the lines to produce the finished drawing. If you work freehand (which is the mode we encourage you to try), draw each line cleanly with a single, careful stroke--don't scribble back and forth. Finally, add notes and labels to explain what each component is.

You may find the exercises difficult at first, but if you follow the procedures we have recommended, they will become easier and more enjoyable as you acquire experience and gain confidence in your growing abilities.

1

MAKING BUILDINGS

Building Code Restrictions

In this exercise you will become familiar with some of the more important ways in which the building code affects the design of buildings. You will need to refer to Figures 1.1, 1.2 and 1.6 of the text, as well as the list of Occupancy Groups provided on page 5, as you do the work. You may also find it helpful to review the example application of these tables to the design of a hypothetical electronics plant beginning on page 10 of the text.

The building code includes many provisions for adjusting height, area, and fire-resistance requirements. For this exercise, apply only the following modifications to the information provided in your text unless directed otherwise by your instructor:

-For buildings two stories in height, the combined area of both floors may be double the allowable area for one floor listed in Table 503 of Figure 1.1.

-For buildings three or more stories in height, the combined area of all floors may be up to three times the area listed in the table.

If the building is fully sprinklered, you may also apply the following adjustments. These adjustments may be applied in combination with those listed above:

-For a single-story building, the allowable building area may be quadrupled.

-For a multi-story building, the allowable height may be increased by 1 story and 20 feet, and its allowable area may be tripled.

1. An old, unsprinklered warehouse of heavy timber construction with exterior walls of brick masonry is being considered for conversion to a drama theater in a small West Virginia town. The building is two stories high, 40 by 70 feet in plan, and conforms to the definition of Type IV (HT) Construction. (Theaters are defined as a Group A-1 Occupancy.)
   a. Will this conversion be permitted? _____
   b. If modifications to exterior bearing walls are required, what fire resistance rating must be provided for this new work? _____

2. A client has asked you to design a clothing store of protected platform frame (Type VA) wood construction. Provide answers for both a sprinklered and unsprinklered building.
   a. What is the maximum total floor area this store can have? _____
   b. How tall can this building be? _____
   c. What is the required fire resistance rating for floor construction? _____

3. What is the maximum height for a reinforced concrete office building of Type 1A construction? _____
   a. What is the required fire resistance rating for a column? _____
   b. What fire resistance is required for floor beams in this building? Why do you think answers in a. and b. differ? _____
   _____
   _____
   _____

   c. If a large concert hall is to be constructed directly abutting the office building, what fire-resistance rating is required for the wall separating these two structures so that they may be treated as separate buildings from a building code standpoint? _____

Name: _____

4. You have decided to use steel framing (Construction Type I or II) for a new five-story hotel (Occupancy R-1) with 41,500 square feet per floor. The building will be fully sprinklered.

a. What is the least expensive (lowest fire-rated) Construction Type you are permitted to use? _____

b. How tall, in feet and number of stories, may the building be? _____

c. What level of fire protection will be required for each of the following elements of this building?
Columns: _____
Floor construction: _____
Roof construction: _____

d. There is a fifteen-foot wide pedestrian passage along one edge of the site that the owner would like to develop as a shopping lane. Can large glass display windows, with a fire-resistance rating of zero, be used along this edge of the building? (See Figure 1.2 in the text, Table 602 of the International Building Code.) _____

5. How tall, in number of stories and in feet above grade, can you build a single-family house (Occupancy R-3) made of wood light frame, unsprinklered, with floor joists and roof rafters left exposed inside (Type VB Construction)?

_____

Real buildings do not get built on paper! Seeing construction take place in the realm of dirt, materials, labor, equipment, and weather is an important part of learning about the making of buildings. The ability to knowledgeably observe work in progress is also an important skill for the design or construction professional. In this exercise, you will visit a construction site to observe work in progress, record your observations, and where necessary, follow up later with analysis of what you have seen.

Since this exercise depends on the constraints of time and access to building sites, your instructor will provide specifics related to the duration and scope of this assignment. It may be performed in the course of a single site visit, or span a series of regular visits to a site over the course of the term. It may involve gaining direct access to a site, or simply observing from a nearby location affording good views of the work in progress.

Observations should be made in the form of notes, and annotated sketches or photographs. In cases where follow-up comments or research are needed, provide concise, clear explanations, and note your sources of information. You may use the form on the following pages as a template for recording your observations and follow-up notes (make additional copies as needed).

During each visit, try to answer as many of the following questions as possible:

1. What **types of work** are underway during your visit--for example concrete pouring, excavation of soil, steel erection, wood framing, etc? Follow-up question: What are the names of the trades performing the work you observed (carpenters doing rough framing, bricklayers laying brick, drywall finishers taping gypsum wallboard, etc.)? For lists of construction trades, review relevant sections in the text. Note that trade designations may also vary regionally.

2. What are the **weather** conditions during your visit (temperature, precipitation, humidity, sky cover)? How is this affecting the work?

3. What **materials** are being stored, delivered, or removed from the site (excavated soil being trucked off-site, delivery of steel concrete reinforcing bars, stockpiling of lumber, etc.)?

4. What are the building's primary **structural materials** (steel frame with cast-in-place concrete floors, light wood frame with OSB sheathing, etc.)? Follow-up: Is this combustible or non-combustible construction? Referring to Figures 1.1 and 1.2 in the text, what Construction Types might this building be?

5. If possible, describe the **exterior wall system**, listing components from exterior cladding to interior finish. Follow-up: For elements that cannot be determined from your observations, suggest possible materials and explain why you think they might be an appropriate choice for this project.

6. What kinds of **temporary supports, construction, or protection** can you see (excavation shoring, erosion control, dewatering, temporary bracing, scaffolding, formwork, tree protection, wind protection, temporary heating, power, safety devices, etc.)? Follow-up: Explain their purpose.

7. What aspects of the **site's physical organization** reflect the need to facilitate the movement of construction materials, labor, and machinery around the site?

8. If you have the opportunity to **talk with a site supervisor**, ask about the organization and challenges of the construction process. How long is the construction planned to take? What activities are most affecting the schedule? What aspects of the construction are most technically challenging or unusual?

9. What do you see that you do not understand? Describe, sketch, or photograph these items. Follow-up: Using the book as a reference or by comparing notes with your classmates, try to explain what you saw.

SITE VISIT REPORT

Project: Date & Time:

Weather: Temp. Range:

Observations & Notes:

Name: _____

It is the rare building project that does not require the contributions of a broad range of participants, including the building owner, architects, engineers, specialized consultants, prime contractors, subcontractors, regulatory officials, user groups, financiers, and more. Achieving a well-built building depends not only on a sound knowledge of construction technology, but also on the ability to communicate effectively and to apply technical knowledge in the context of a project's often competing priorities and complex web of participants.

This exercise is unlike any other in this workbook. Its focus is on communication and teamwork rather than building techniques and materials. You will form a group representing key players in the building process. Your goal is to complete a simple construction project, from initial conception to finished product. Don't be fooled by the seemingly simplistic nature of the construction itself. In this exercise, we are deliberately choosing a technology with which almost every student of design and construction is familiar: paper and glue!

You should gain from this exercise an appreciation of the challenges in achieving a coherent and successful project in the context of a process that involves many participants. When you have finished this exercise, imagine increasing the scale of complexity many orders of magnitude, as is the case with almost any real-life project. As you proceed through the remainder of these exercises and your course work, remember that successful building construction requires both technical knowledge and the skills to apply that knowledge effectively.

Good luck!

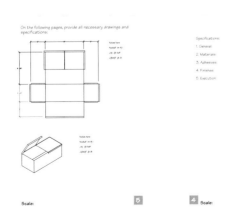

In this exercise you will form a project team, and design and build a paper object within a limited budget. The constructed object is to be made only from paper of any weight, and glue.

1. Team up with three other classmates, and choose among yourselves the roles of Owner, Designer, Consultant, and Builder.

2. The Owner is to write a concise project statement describing broadly the goals for the project. What kind of object is desired, how should it look? Don't try and describe how it is made or define its characteristics in detail. For example, "I would like a portable box to hold my drafting tools. It should be sufficiently durable to last the semester..."

3. The Owner, Architect, and Builder are to meet and review the project statement. All three parties must agree on a time limit for both the design and construction phases of the project. If necessary, changes in the Owner's requirements should be negotiated until all parties are satisfied that the project is achievable within acceptable limits.

4. The Architect is to meet next with the Consultant. The Architect and Consultant will prepare the construction documents, consisting of drawings and a written specification. The drawings should describe the shape, size, and arrangement of the object and its parts. The specification should describe the materials and quality of construction, and provide any necessary assembly or finishing instructions. It is up to the Architect and Consultant to organize their efforts so as to efficiently combine their efforts and produce the required documents within the established design budget.

5. At an intermediate point during the Architect and Consultants' work, all team members are to meet to review the design in progress and ensure that the Owner's and Builder's requirements are being satisfactorily addressed.

6. Copies of the construction documents are to be delivered to the Builder, who is to execute the documents and construct the finished object. The Builder also is obligated to complete the work within the established budget for construction.

7. While design and construction are underway, the Owner is to build a version of the object as well, based on the Project Statement, but without relying on the Architect and Consultants' construction documents.

8. After both constructed objects are completed, the team should meet as a group to consider the following questions:

   a. How successfully was the Owner's original intent achieved in the final product?

   b. How did the Owner's version differ frm the team's--was one or the other more successful at fulfilling the original project statement?

   c. How did the division of labor among the project team help to improve the results of the project? Did different, contributing points of view lead to a better design? Did the "checks and balances" of a team help reduced errors?

   d. How did the team approach hinder a satisfactory outcome? Did the Owner's goal get lost in the translation? What kinds of misunderstandings occurred? Did different team members have conflicting goals for the project?

   e. How do you imagine these issues playing out in real-life design and construction projects?

On the following pages, provide all necessary drawings and specifications. Additional pages may be added as necessary.

Owner: _____
Architect: _____
Consultant: _____
Builder: _____   Scale:

Scale:

14 Scale:

2

FOUNDATIONS

Waterproofing and Drainage

The materials used to protect foundations from moisture are referred to as either "dampproofing" or "waterproofing". Dampproofing materials are water-resistant, but not adequate for resisting the passage of water under hydstatic pressure. Where drainage conditions are poor, or ground water may be present, materials classified as waterproofing are recommended.

Where waterproofing is required, the choice of system can depend on a variety of factors. Here are a few examples:

a. **Liquid-applied** membranes that cure in place are relatively easy to detail around complex shapes and penetrations, since in liquid state, they can be easily formed to any shape.

b. Sheet membranes that are **loosely laid**, rather than fully adhered, are well-suited for use over substrates prone to movement or cracking, since movement in the substrate is less likely to transmit stress into the membrane.

c. Membranes that are **fully adhered** to the substrate may better limit leakage caused by a minor defects in the membrane, since they are less likely to permit water to travel under the membrane and spread to areas remote from the origin of the leak.

d. Most foundation waterproofing systems must be applied to the exterior side of the foundation wall. **Cementitious** waterproofing, made by the addition of waterproofing agents to portland cement plaster, bonds well enough to concrete to allow its application on the inside of a concrete wall that is exposed to water on its exterior.

e. Many waterproofing systems can only be applied over a dry substrate. **Bentonite clay** is one example of a waterproofing material that can be applied over uncured concrete, potentially an advantage when construction takes place during extended periods of cold and damp.

1. For each condition below, indicate whether dampproofing or waterproofing is most appropriate:

   a. Below-grade space for housing library stacks

   _____

   b. Crawlspace in well-drained soil

   _____

   c. Below-grade utility room, in normally-drained soil

   _____

   d. Finished basement, in normally-drained soil, where owner has expressed particular concerns regarding moisture damage and mold growth

   _____

2. For each of the following, propose a waterproofing system and comment briefly on the reason for your choice:

   a. A concrete basement poured in the winter, which is likely to remain damp for many months.

   b. A concrete foundation carrying a prestressed concrete deck. The deck is likely to creep and cause significant cracking in the foundation wall over an extended period.

   c. A concrete elevator pit below grade. The exterior sides of the pit are cast directly against the excavation and will never be accessible for application of waterproofing.

   d. A foundation for an underground mechanical room. The foundation is geometrically complex, and is penetrated in many places to permit the passage of pipes and wiring conduits.

Name: _____

3. Complete the following foundation section to include a waterproof membrane on the exterior of the wall, insulation, a drainage system, backfill, and finish grade. Label all features contributing to waterproofing. For guidance, refer to Figures 2.55, 2.57, and 2.60 of the text.

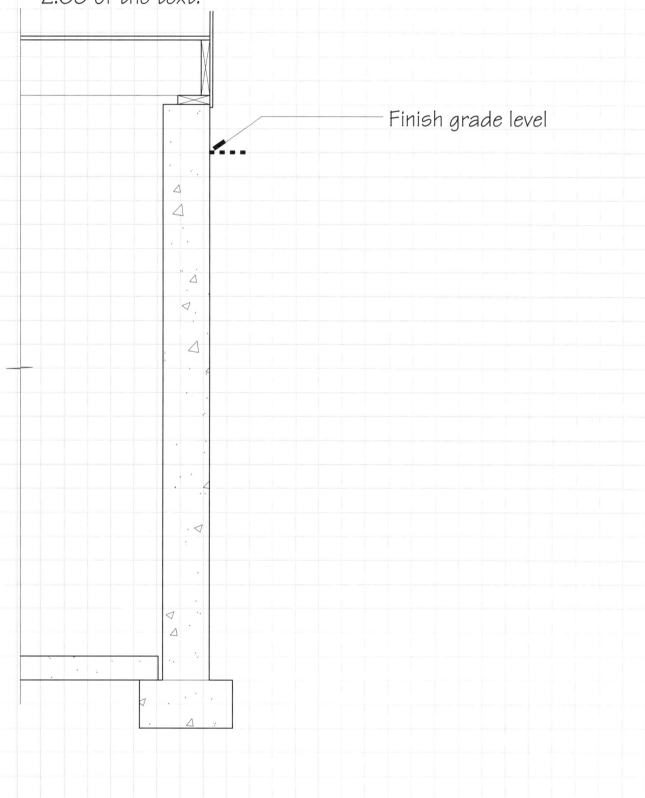

Finish grade level

Scale: 3/4" = 1' (1:16)
1 square = 4" (100 mm)

For assistance with this exercise, refer to Figures 2.2 and 2.5 of the text.

1. Give one or two possible identifications for each of the following. Provide a Group Symbol and descriptive name for each. It is not necessary to distinguish well-graded from poorly-graded soils:

   a. All of the soil particles are visible. Some of the particles are large enough to be picked up individually, but most cannot.

   b. When dry, the soil seems to be a dusty sand. When wetted it is still gritty like sand, but the soil sticks together in a ball if compressed in the hand.

   c. No individual soil particles are discernible by eye, but the soil came out of the ground in hard chunks. When a small sample is wetted it becomes a sticky paste that can easily be molded into shapes.

   d. The smallest particles in the soil can be individually lifted between two fingers, the largest with the whole hand.

   e. No soil particles are discernible by eye, yet the soil, even when wet, falls apart when an attempt is made to mold it into a shape.

   f. The soil smells musty and is very dark in color. It seems to spring back slightly after being compressed in the hand.

2. Which of the above soils is likely to have the highest loadbearing capacity under a wall footing or strip footing?

3. Which of the above soils would you expect to drain freely?

Name: _____

4. How large does a square column footing need to be to support a load of 85,000 pounds (39,000 kg) on a compact sandy gravel soil? Show calculations. Make a sketch of the footing, assuming that it is 12" (300 mm) thick.

5. How wide must a wall footing be if the load is 3,200 lb (21,000 kg) per foot of wall length, and the footing rests on a sandy clay soil? Show calculations and make a sketch. Assume the footing is 12" (300 mm) thick.

Foundation and Slope Support Systems 2.3

1. Three excavations are shown below in cross section. Draw a slope support system for each as indicated.

a. Steel sheet piling supported by cross-lot bracing made of steel wide-flange shapes (see Figure 11.10 of the text for illustration of wide flange shapes)

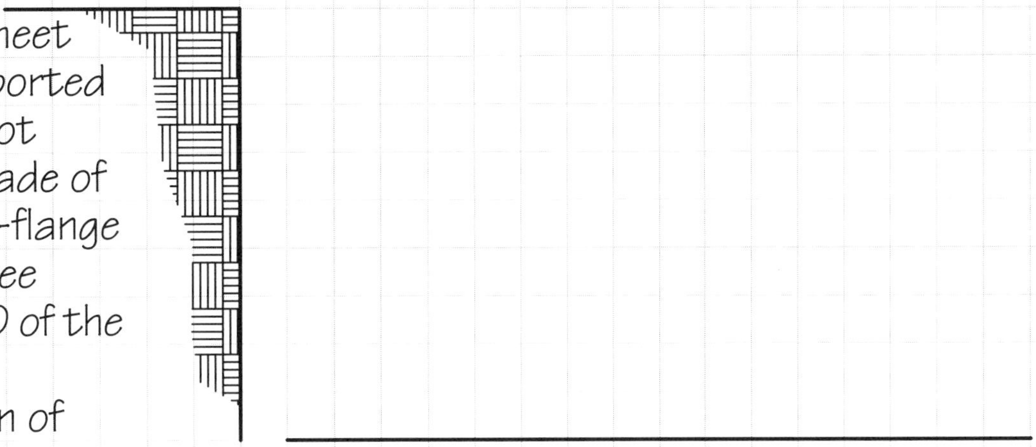

b. Soldier beams and wood plank lagging supported by heavy timber rakers

c. Slurry wall supported by tiebacks

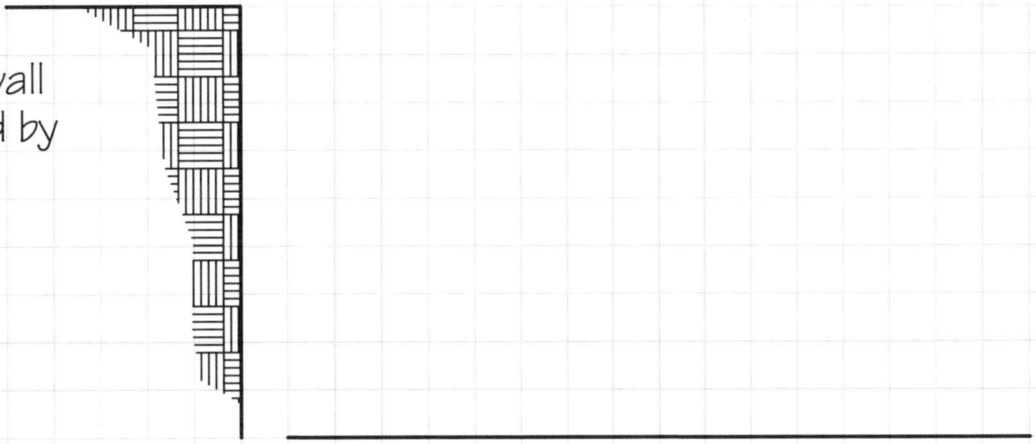

Name: ______________________ Scale: 1/8" = 1" (1:96) 1 square = 2' (600 mm)

2. On the section below, draw foundation elements as indicated.

    a. Caisson 4' (1200 mm) in diameter with a bell 2.5 times this diameter

    b. Precast concrete end bearing pile, 16" (400 mm) square

    c. Concrete bearing wall 12" (300 mm) thick supported by a strip footing 36" wide and 12" deep (900 x 300 mm) resting on soil 48" (1200 mm) below grade

    d. Cluster of 16 wood friction piles each 30' (9 m) long with an average diameter of 12" (300 mm), spaced 36" (900 mm) apart

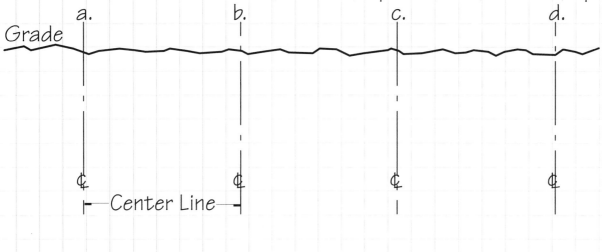

Grade

a.   b.   c.   d.

¢   ¢   ¢   ¢

⊢—Center Line—⊣

Firm bearing stratum

22

Scale: 1/8" = 1' (1:96)
1 square = 2' (600 mm)

3

WOOD

The next two exercises will help you begin to find your way toward a good working knowledge of wood, wood products, and wood fasteners. You need to apply facts and figures from many parts of Chapter 3 of the text, so the work will go more smoothly if you review the chapter thoroughly before you begin.

1. Name two wood species appropriate for each of the following uses:
   a. Hardwood flooring _____ _____
   b. House framing _____ _____
   c. Window and door frames _____ _____
   d. An outdoor deck _____ _____
   e. Softwood flooring _____ _____
   f. Fine furniture and paneling _____ _____

2. Circle the end uses below for which quartersawn lumber is preferred:

   framing members   interior trim   sheathing boards
   finish flooring   outdoor decks   tabletops

3. Give actual dimensions (in English and metric units) for each of the following pieces of lumber:
   a. 2x4 _____
   b. 2x6 _____
   c. 2x8 _____
   d. 2x12 _____
   e. 1x4 _____
   f. 1x12 _____
   g. 4x6 _____
   h. 6x6 _____

4. What is dimension "x" (in both English and metric units) in the detail below? Show calculations.

5/8" plywood
2x10
2x6
X

Name: _____

5. A board exactly 12" (305 mm) wide was quartersawn from a green softwood log, then seasoned to a moisture content of 12%. How wide is it now? Show calculations.

6. The platform frame shown in Figure 5.2 of the text contains a total of 33" (838 mm) of cross-grain wood between foundation and roof.
   a. Assuming that plainsawed framing lumber shrinks across its grain at a rate that is an average of the shrinkage rates of tangential and radial shrinkage, how much will the roof drop if the lumber is installed at 19% moisture content and eventually dries to 15%?
   b. Assuming that the 2x12 wood floor joists at both floor levels are replaced with laminated veneer lumber joists with negligible shrinkage, how much will the roof drop under the same change in moisture conditions?
   Show calculations.

7. Considering the tendency of plainsawed lumber to cup during seasoning, which way should the boards on an outdoor deck be laid, so that they will not trap rainwater?
   a. Circle the properly laid board.

Top of supporting beam

   b. Can you think of any other factors that might influence the choice of orientation?

1. How many board feet are contained in a 2x4 stud 8' long? If the stud costs $1.33, what is the cost per board foot? Show calculations.

2. What will be the cost of 34 2x10 floor joists, each 12' long, if the price of lumber is $330.00 per thousand board feet? Show calculations.

3. List a softwood plywood veneer grade suitable and economical for each of the following uses:
    a. Reverse side of a low-cost plywood panel that will not be seen _____
    b. Painted face of a storage cupboard _____
    c. Smooth but low-cost floor surface over which carpet will be laid _____

4. List a softwood plywood exposure durability classification suitable for each of the following proposed uses:
    a. Structural sheathing, subflooring _____
    b. Exterior siding _____
    c. Item c. in question 3 above, not to be installed until the building is enclosed _____

Name: _____

5. Explain a "32/16" span rating stamped on a sheet of plywood.

6. What size common nail (designated in pennies) will just reach completely through two thicknesses of nominal 2-inch lumber?

_____

7. If 5/8" oriented strand board sheathing is nailed to the wood frame of a building with 8d common nails, how far does the point of the nail penetrate into the frame?

_____

8. How far does a 16d nail end penetrate into a longitudinal piece of wood after it has been driven through a nominal 2-inch piece?

_____

9. Match the following nails with their uses:
   a. Finish nail   attaching wood shingles
   b. Deformed shank nail   attaching interior wood trim
   c. Cut nail   attaching gypsum wallboard
   d. Box nail   attaching asphalt shingles
   e. Roofing nail   attaching hardwood flooring

10. Match each of the following composite wood products with its description:
   a. laminated veneer   large flakes of wood compressed and
      lumber   bonded into sheets
   b. parallel strand   veneer sheets laminated into
      lumber   rectangular sections
   c. plywood   small wood particles, compressed and
   d. composite panel   bonded without orientation into
   e. oriented strand   panels
      board   thin wood veneers, glued into panels
   f. waferboard   two parallel face veneers bonded to
   g. particleboard   a reconstituted wood fiber core
                              narrow veneer strands, oriented
                                 longitudinally and pressed into
                                 rectangular cross sections
                              long, strand-like wood particles,
                                 compressed and glued into sheets

28

HEAVY TIMBER FRAME CONSTRUCTION

4.1 Heavy Timber
Framing

Exercise 4.1 appears simple, but deserves consideration of several alternative solutions before one is selected. You have many options: You can use thicker decking and space the beams farther apart, or thinner decking and more closely spaced beams. Similarly, you can use larger beams and space the girders farther apart, and by using larger girders you can space the columns more widely. Experiment with different framing plans on scratch paper, then choose one that seems to you to consist of a balanced set of components with reasonable sizes and spacings.

Some general guidelines for this exercise: Use decking that is nominally 2", 3", or 4" deep (38, 64, or 89 mm). Support the decking with beams, and the beams with girders. If you divide the building into bays that are not square, span the longer dimension with the girders. Use the structural rules of thumb on page 135 of the text to arrive at approximate member sizes. Remember also to abide by the <u>minimum</u> member size restrictions for Heavy Timber construction given in Figure 4.7 of the text.

You may use solid wood or glue-laminated members. Solid wood beams and girders should be no deeper than 24" (600 mm). For standard sizes of laminated members, refer back to page 91 of the text, keeping in mind that member depths should be some multiple of a single lamination thickness, typically 1 1/2" (38 mm).

Base your connections on any details in Chapter 4 that seem appropriate. If you choose to use cantilevered beams joined with hinge connectors as shown in Figure 4.15, locate the hinge connectors at a distance from the column approximately 1/5 of the total distance between the columns.

1. Shown below is the floor plan of a two-story furniture factory in Idaho. The exterior wall is made of 12" thick (300 mm) concrete blocks. Draw a framing plan for the upper floor of this building, using a construction of timber decking supported on laminated wood beams, girders, and columns. Indicate approximate sizes of all members.

Name: ____________________ Scale: 1/16" = 1' (1:192) 1 square = 4' (1.2 m)

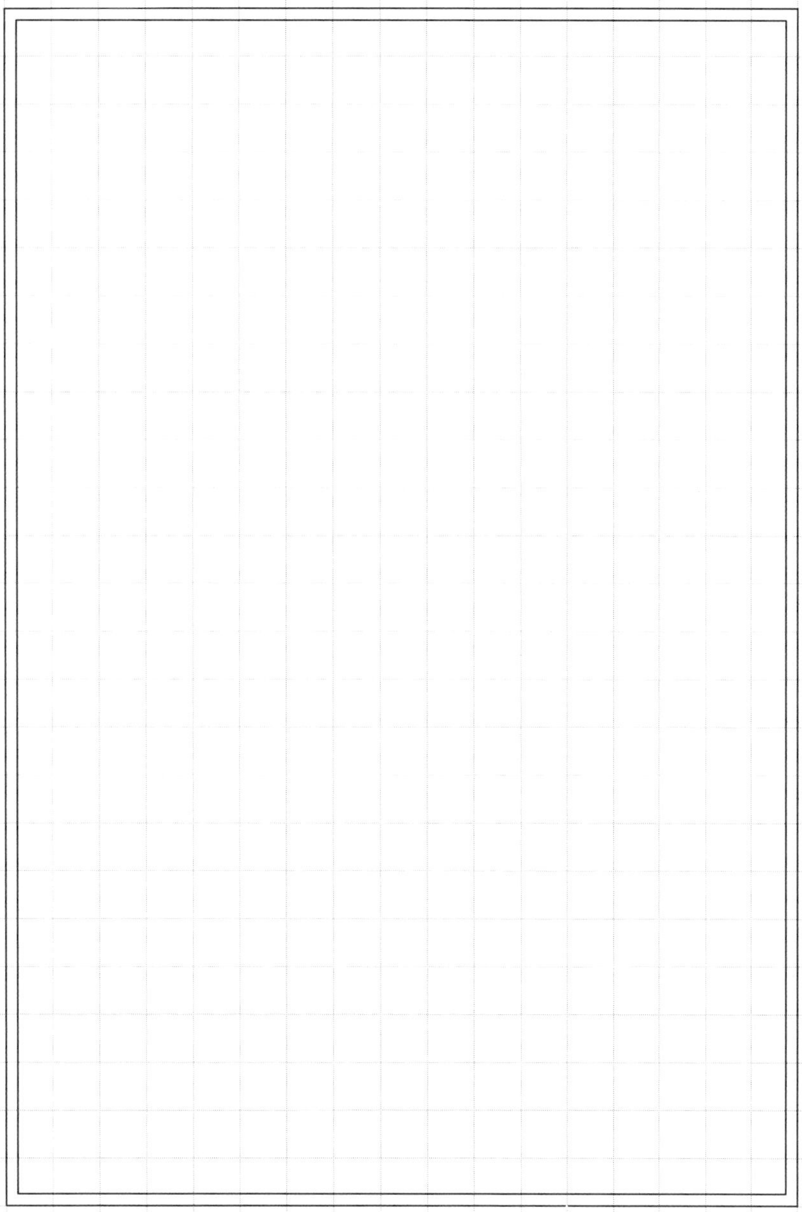

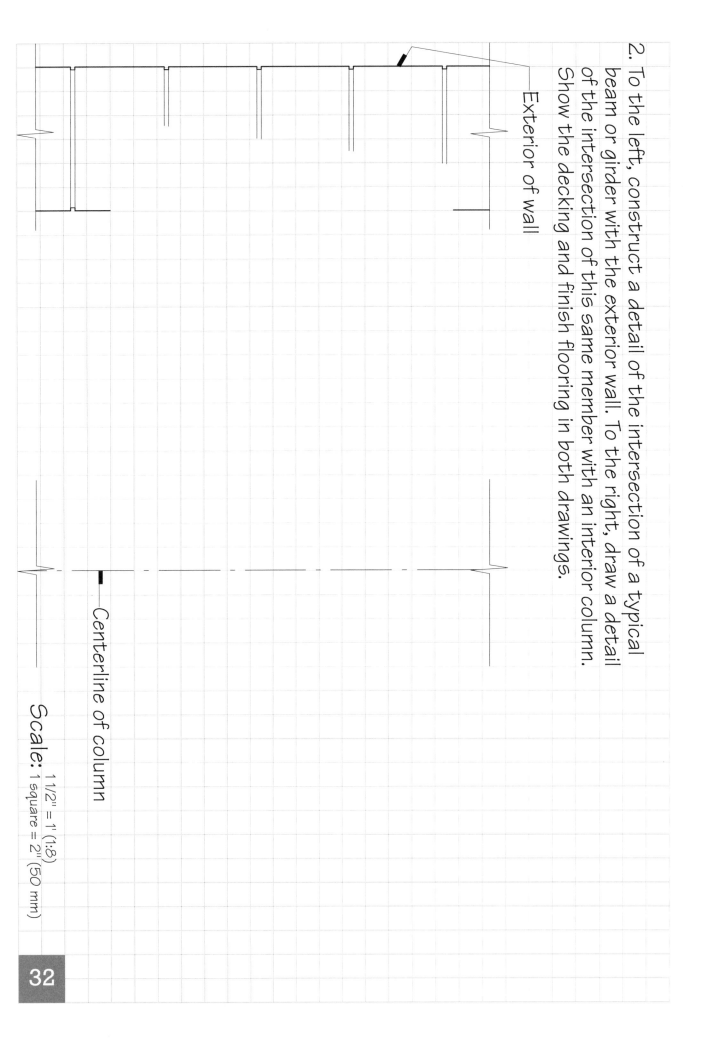

2. To the left, construct a detail of the intersection of a typical beam or girder with the exterior wall. To the right, draw a detail of the intersection of this same member with an interior column. Show the decking and finish flooring in both drawings.

Exterior of wall

Centerline of column

Scale: 1 1/2" = 1' (1:8)
1 square = 2" (50 mm)

5

WOOD LIGHT FRAME CONSTRUCTION

1. Review Figures 5.17 through 5.20 and pages 163 through 165 in your text. Referring to the preliminary design guideines on page 189, jot down the maximum spans for 2x8, 2x10 and 2x12 wood joists. (See the note below if you wish to complete this exercise with I-joists.)

2. Examine the floor plan, searching for a simple arrangement of joists and beams, working within the span limits noted above. To avoid complications for the carpenters, use one size of joist throughout.

4. Draw in the beams and locate posts assuming beams can span 15' to 20' (4.6 m to 6.1 m) between supports. Add doubled headers and trimmers around stairs, chimneys, and other floor openings. Add doubled joists wherever partitions run parallel to the framing below.

5. Next lay out the framing for any cantilevered bays. The length of a cantilever should be no more than one-third the length of the interior span, and not more than one-fifth of the allowable span for the joist.

6. Complete the framing plan by adding the remainder of the joists at a spacing of 16" (400 mm) o.c. Start at one edge and work across the platform, faithfully maintaining this spacing regardless of the placement of other framing members.

7. Add joist hangers wherever joists are supported by headers. Add solid blocking wherever joists span across a beam or cantilever over a wall. Consider adding bridging at the midpoint of longer spans or deeper joists.

You may also complete this assignment using I-joists as follows: Limit the maximum I-joist depth to 12" (300 mm); space I-joists at 24" (600 mm) o.c.; substitute structural composite beams, such as LVLs, for doubled joists and headers; and limit cantilevers to one-fourth of the interior span.

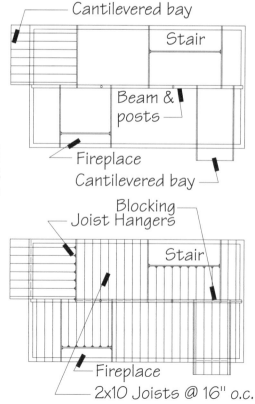

Cantilevered bay

Stair

Beam & posts

Fireplace

Cantilevered bay

Blocking

Joist Hangers

Stair

Fireplace

2x10 Joists @ 16" o.c.

1. Design and draw ground floor framing layouts for the buildings whose foundation plans are shown on this page and the following. Designate the size and spacing of joist used in each, and show beams, posts, doubled joists, joist hangers, blocking, and other features.

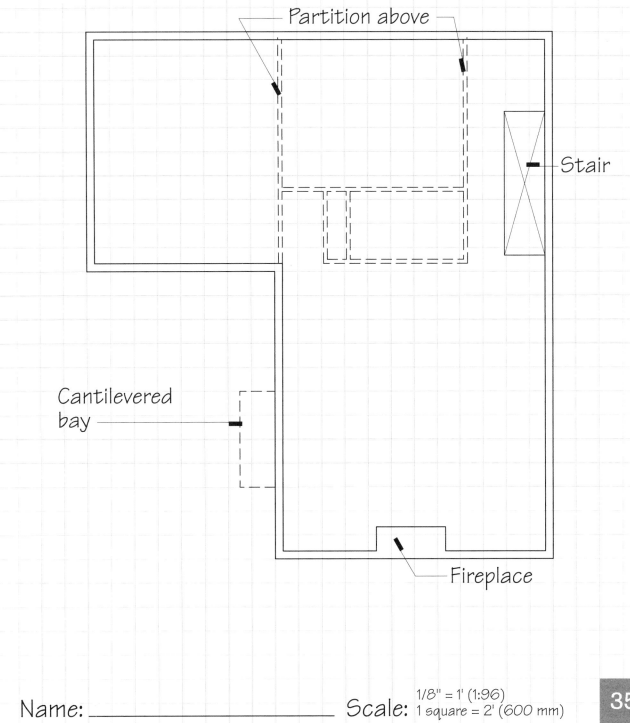

Partition above

Stair

Cantilevered bay

Fireplace

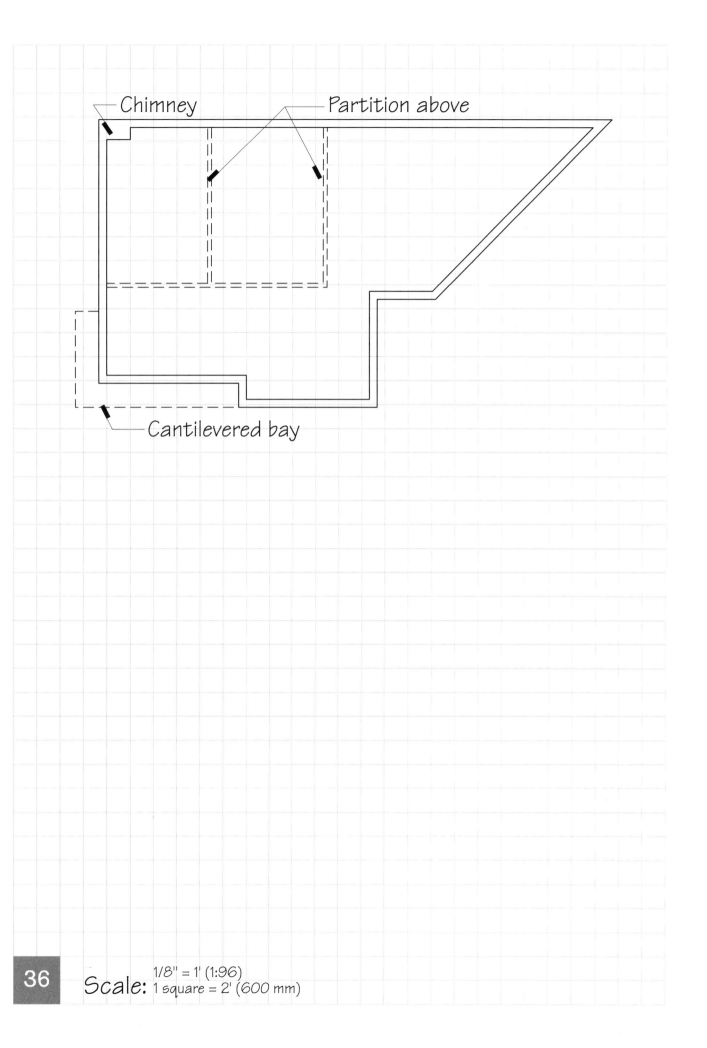

Chimney

Partition above

Cantilevered bay

Scale: 1/8" = 1' (1:96)
1 square = 2' (600 mm)

1. Design and draw complete framing for these two exterior walls, using 2x6 studs 24" o.c. (38 x 140 mm studs 600 mm o.c.) and 1x4 (19 x 89 mm) let-in diagonal bracing. Follow the procedure outlined in Figure 5.32 of the text. For door and window headers, use 2x8 (38 x 184 mm) framing or larger. Lettered arrows refer to details on the following page.

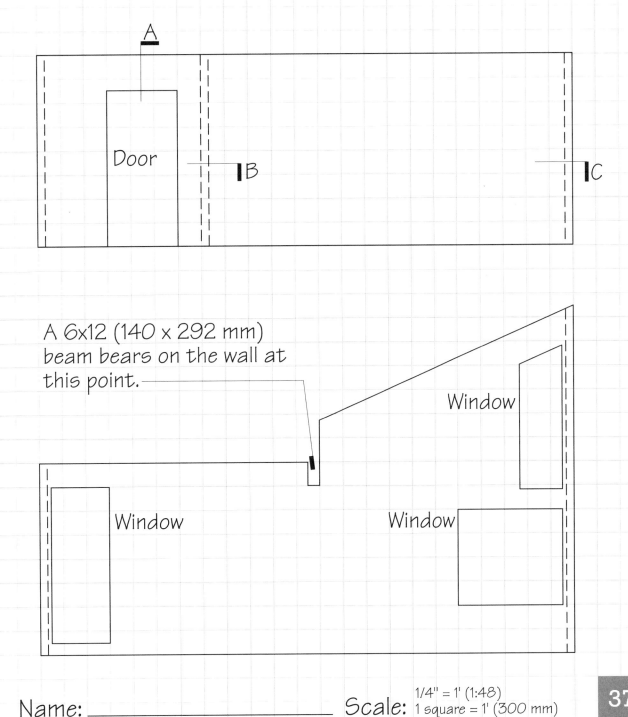

A 6x12 (140 x 292 mm) beam bears on the wall at this point.

2. Draw details A, B, C from the previous page in the space below. See Figures 7.18 and 7.19 in the text for 2 x 6 framing detail examples.

Top of wall

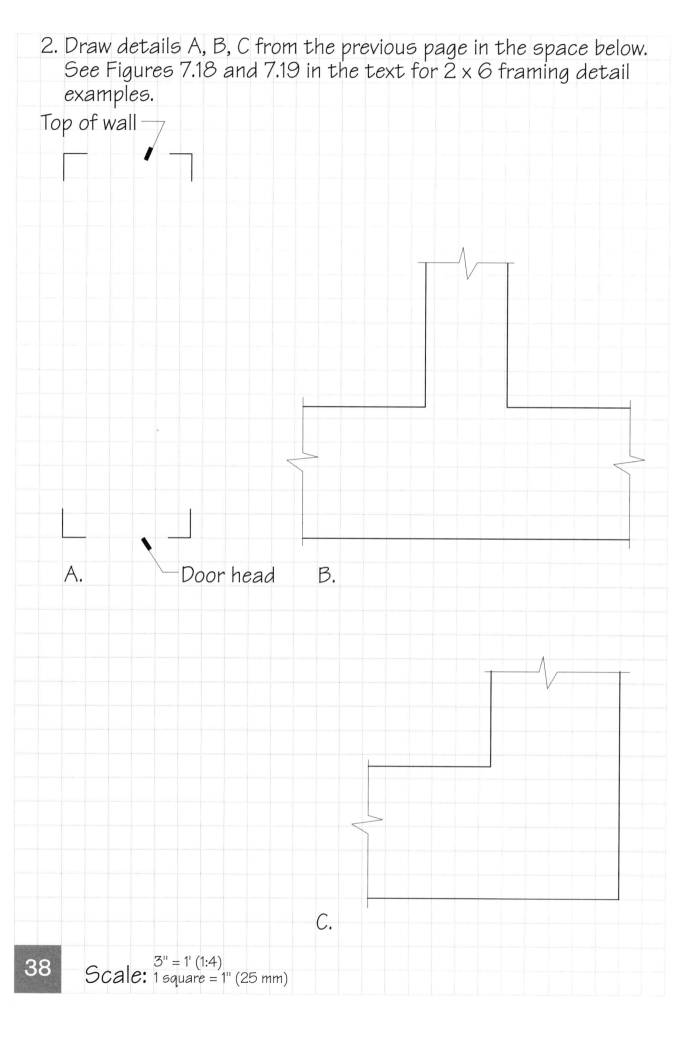

A.

Door head   B.

C.

Scale: 3" = 1' (1:4)
1 square = 1" (25 mm)

The basic building block of all pitched roof configurations is the shed, or single-pitched roof:

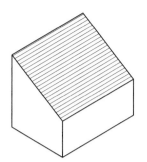

Two sheds together make a gable roof:

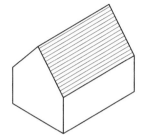

Two intersecting gables make a hip roof:

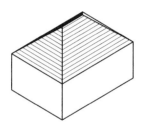

Or a dormer:

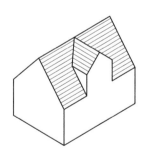

Pitched roofs can be added together to shelter almost any collection of interior spaces.

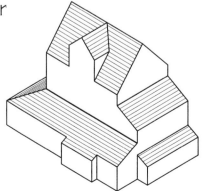

(continued from previous page)
The easiest way to do calculations with roof pitches is to set up a proportion using the given rise and run:

Find the height "y" of the roof at a distance of 7'-6" (2285 mm) from the edge.

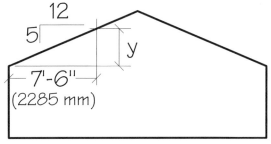

Solution: $\dfrac{5}{12} = \dfrac{y}{7.5'}$

$y = \dfrac{5\,(7.5')}{12}$

$y = 3.125' = 3'\text{-}1\,1/2"$

At what horizontal distance from the eave will this roof have risen six feet (1830 mm)?

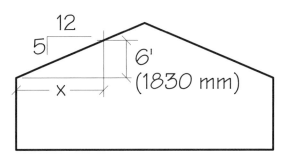

Solution: $\dfrac{x}{6'} = \dfrac{12}{5}$

$x = \dfrac{6'\,(12)}{5}$

$x = 14.4' \cong 14'\text{-}4\,1/2"$

1. Study the two examples below, then draw roof plans and thumbnail perspectives of ten more ways of covering an L-shaped building with roofs at a pitch of 8/12.

2. You are remodeling a 23-foot wide (7010 mm) attic beneath a roof at 8/12 pitch. How wide a room can you create between new walls that are 5'-0" (1524 mm) high? Show calculations.

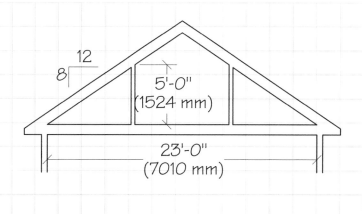

3. If you made the room in the problem above 11' (3353 mm) wide, how high would the walls be? Show calculations.

4. What is the widest balcony that can be built in the building shown to the right while maintaining a minimum headroom above the balcony of 6'-6" (1420 mm)? Show calculations.

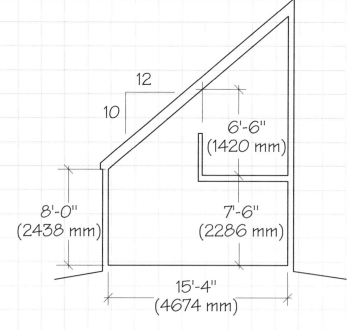

To the right is the ground floor plan of a vacation cottage. The Dining Area and Kitchen should have a ceiling height of 7'-6" (2286 mm). The Loft floor is 12" (305 mm) thick, and the Loft should have a minimum headroom of 6'-8" (2032 mm). The Living Room should be a dramatic room that rises to the sloping underside of the roof framing.

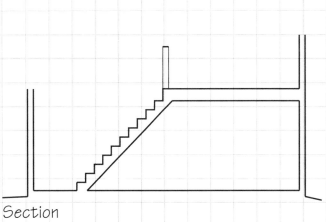

Fireplace — Loft Above

Living Room   Dining Area

Up

Bath

Floor Plan

1. In the space below, draw a small freehand perspective of a design for a roof for this cottage. The roof may consist of any number of planes, but all must slope at the same pitch, which may not be less than 6/12.

Section

2. Complete the section drawing above to indicate how your roof design looks when cut at the section line indicated on the Floor Plan. Show the numerical pitch of the roof and give all wall heights and headroom clearances.

Name: _____

Scale: 1/8" = 1' (1:96)
1 square = 2' (600 mm)

3. Complete the roof framing plan below in accordance with your roof design. Use rafters spaced 24" (600 mm) o.c. Assume that the rafters can span no more than 12'-0" (3660 mm) in plan, and that one or more wood beams, as indicated on your plan, may be used to support rafters. No rafters may rest on the chimney.

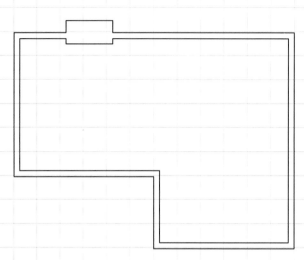

Roof Framing Plan

Scale: 1/8" = 1' (1:96)
1 square = 2' (600 mm)

EXTERIOR FINISHES FOR WOOD LIGHT FRAME CONSTRUCTION

6.1 Exterior Detailing

Exterior Detailing

This exercise requires that you bring together the various details of exterior finishes from Chapter 6 of the text and apply them in a consistent manner to a single building. You will find Figures 6.1, 6.2, 6.14, 6.15, 6.22-6.24, and 6.28 to be particularly helpful.

1. To this exterior wall section of a small dental clinic, add thermal insulation, roof ventilation features, and a set of materials of your choice, to create a completely finished exterior. No gutters will be installed. The rafter ends are to be cut as shown, with the fascia and soffit attached directly to them. Label all materials.

2x6 Studs

2x10 Joists

Name: _____

Scale: 1 1/2" = 1' (1:8)
1 square = 2" (50 mm)

2. For the same building and materials, complete and label the details shown below.

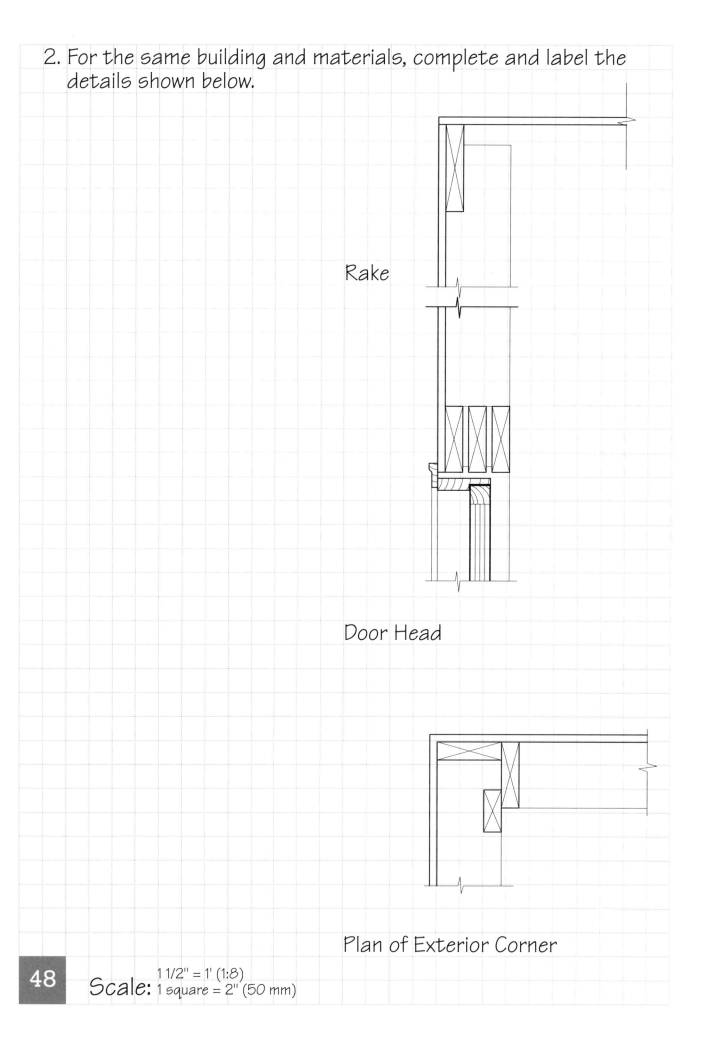

Rake

Door Head

Plan of Exterior Corner

Scale: 1 1/2" = 1' (1:8)
1 square = 2" (50 mm)

7

INTERIOR FINISHES FOR WOOD LIGHT FRAME CONSTRUCTION

Pages 246 through 248 of the text give general guidelines and precise dimensions for proportioning conventional masonry fireplaces.

1. Your teacher will assign a fireplace opening width. Starting from this dimension, complete the three views below and show all the critical dimensions for a correctly proportioned conventional fireplace.

Floor Level

Section

Elevation

Plan

Flue lining dimensions:

_____

2. Make a freehand three-dimensional drawing in the space below to show how this fireplace and hearth will look. You may use any masonry materials, tiles, and/or mantel materials you wish.

Proficiency with laying out stairs and proportioning their treads and risers, even at times under seemingly impossible circumstances, is an indispensable skill for the designer of buildings. For this exercise, review pages 258 through 259 of the text for the information you will need.

1. Calculate numbers and dimensions of treads and risers for the following stairs. Show all calculations.

a. Exit stair in a high school, total rise 12'-8" (386 mm).

Number of risers: _____
Riser height: _____
Number of treads: _____
Depth of tread: _____

b. Main stair in a single-family residence, total rise 8'-11 1/2" (2731 mm).

Number of risers: _____
Riser height: _____
Number of treads: _____
Depth of tread: _____

c. Exterior entrance steps to a courthouse, total rise 4'-9" (1448 mm).

Number of risers: _____
Riser height: _____
Number of treads: _____
Depth of tread: _____

Name: _____

2. You are designing the renovation of a beautiful 70-year-old house. The plan of the entrance room is shown below. You must add to this room a new stair to an upper floor that lies 9'-2" (2794 mm) above, and the owner requests that it be 3'-6" (1067 mm) wide. Work out the tread and riser proportions for this stair, and draw your design for the stair carefully to scale on the plan drawing. Show calculations.

Number of risers: _____
Riser height: _____
Number of treads: _____
Depth of tread: _____

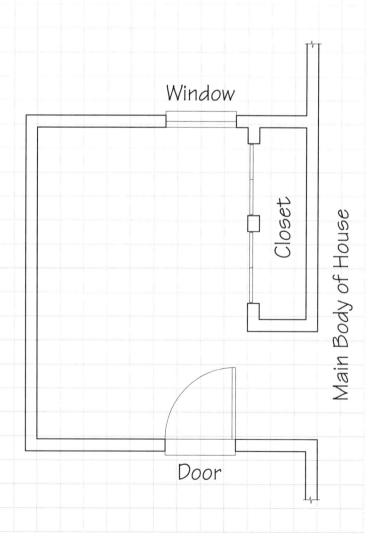

Window

Closet

Main Body of House

Door

Scale: 1/4" = 1' (1:48)
1 square = 1" (300 mm)

Two friends of yours operate a successful business conducting bicycling tours through the wooded hill country of southern Wisconsin. They would now like to build a half-dozen simple hostels on scattered rural roadside sites to serve as overnight shelters for tour groups, and have asked you to design a prototype. You have agreed with them that each hostel should be a single-story wood platform frame building of about 750 square feet (70 m²) with a sleeping loft above that is about half this size. The main floor should have a single bathroom with toilet, wash basin, and shower, a rudimentary kitchen for heating prepared meals, and a fireplace. Outside, an attached rain shelter for up to 20 bicycles is required. Materials for exterior and interior should be simple, rugged, and consistent with the rural settings of the hostels.

1. Draw a floor plan of your design at the indicated scale.

2. Draw a small perspective illustrating the exterior appearance of the hostel.

3. Draw an accurate roof framing plan at the same scale as the floor plan on the previous page.

Scale: 1/8" = 1" (1:96)
1 square = 2" (600 mm)

4. List materials for the following uses, specifying species, grade and/or method of sawing as appropriate.

    a. Foundation type and material:

    b. Joists and rafters:

    c. Studs:

    d. Subflooring:

    e. Sheathing:

    f. Roofing:

    g. Siding:

    h. Fascias, corner boards, exterior trim:

    i. Exterior doors:

    j. Window type and material:

    k. Thermal insulation:

    l. Ceiling finish material:

    m. Wall finish material:

    n. Flooring and stair treads:

    o. Interior trim:

    p. Entrance stairs and deck:

5. Draw a complete eave detail reflecting your design for the hostel as shown in the perspective. Include all interior and exterior finishes, using the materials you have specified. Label all components.

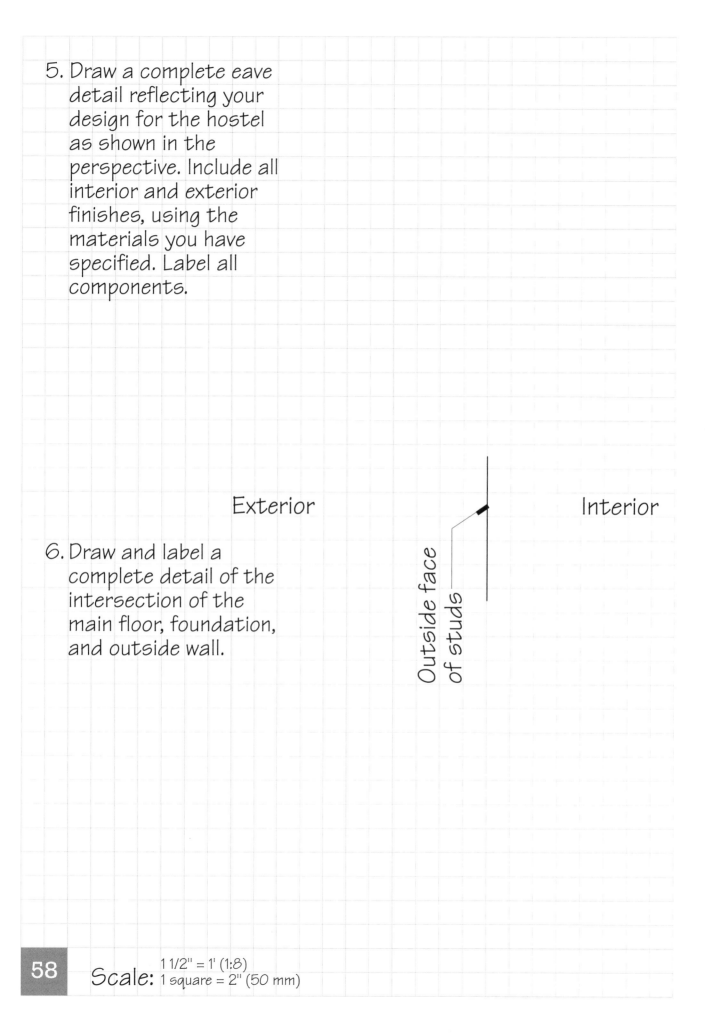

Exterior

Interior

Outside face of studs

6. Draw and label a complete detail of the intersection of the main floor, foundation, and outside wall.

Scale: 1 1/2" = 1' (1:8)
1 square = 2" (50 mm)

8

BRICK MASONRY

Selecting Bricks and Mortar

The selection of bricks and mortar for a building project is based on both aesthetic and functional considerations. For the first two problems in this exercise, use Figures 8.5, 8.6, 8.15, and 8.16, and the accompanying text, to make brick and mortar selections for projects with regard to strength, weather resistance, and appearance requirements. For problem 3, consult Figure 8.11 for information on the sizes of various bricks.

Following are some additional materials selection guidelines:

a. The selection of **brick grade** is based on resistance to weathering, and especially freeze-thaw action. For reasons of economy, choose the lowest acceptable grade.

b. The selection of **brick type** is based on appearance considerations. There is not necessarily a correlation between type and quality or cost. Often, bricks that are more nonuniform are prized for the color patterns and textures they create in the wall. Compared to more uniform brick types, they may be more costly to manufacture or more difficult to lay up.

c. The selection of **mortar type and mix** requires balancing considerations of strength, freeze-thaw resistance, workability, compatibility with the brick units, and cost. As a general rule, select the weakest mortar type that is suitable for the job. For example, where masonry is in contact with the ground, Type M mortar is the recommended choice for its durability and resistance to frost action. For exterior reinforced or loadbearing masonry, Type S mortar is recommended, and for brick veneer and other nonloadbearing exterior work, Type N. Type O mortar is frequently specified for masonry restoration work, where a stronger, harder mortar would risk overstressing older, softer brick units.

1. Indicate appropriate brick and mortar selections for each of the following:

    a. Little Rock, Arkansas: Exterior loadbearing walls for a 17-story dormitory with a highly-regular and smooth appearance

    b. Winnipeg, Manitoba: Wall facing inside a shopping mall, with a "hand-made brick" look

    c. Palm Springs, California: Brick retaining wall with a very rough texture

    d. Cody, Wyoming: Exterior brick facing on a single-story house, with a moderate range of size and color variation

    e. Mobile, Alabama: Smooth brick sidewalk

    f. Seattle, Washington: Variegated bricks for an exterior two-story loadbearing wall

| | Bricks | | Mortar |
	Grade	Type	Type

2. Give two alternative recipes for type S mortar:

a.	b.

3. Starting with a 3/8" (9.5 mm) mortar joint at the bottom, fill each rectangle with an elevation view of running bond brick work using the designated size of brick. Draw each brick and its surrounding mortar joints as accurately as you can.

Modular

Standard

Draw your hand to the same scale here

Roman

King Size, 2 5/8" high

Scale: 1 1/2" = 1' (1:8)
1 square = 2" (50 mm)

In the spaces below, draw elevations and corresponding cross sections of brick walls in each of the indicated bonds. Use modular bricks and a nominal 8" (200 mm) wall thickness. Draw mortar joint thickness accurately.

Outside corner of two intersecting walls

English Bond, weathered joint--Elevation   Section

Flemish Bond, raked joint--Elevation   Section

Common Bond, concave joint--Elevation   Section

Name: _____   Scale: 1 1/2" = 1' (1:8)
1 square = 2" (50 mm)

63

Another bond of your choice or invention--
Elevation

Section

Scale: 1 1/2" = 1' (1:8)
1 square = 2" (50 mm)

If you study Figure 8.22 of the text, you will see that the derivation of building dimensions from masonry unit dimensions is not difficult, but does require some care.

The basic unit of horizontal dimensioning for a brick building is the length of one brick plus the thickness of one mortar joint. If we are using a modular brick, this unit is 7 5/8" for the brick plus 3/8" for the mortar joint, which add up to 8" (194 mm plus 9 mm equals 203 mm). But a masonry wall that stretches from one outside corner to another outside corner always has one fewer mortar joints than bricks--looking at the left hand portion of the wall in the figure, you will be able to count 7 bricks and 6 mortar joints. The easiest way to figure the exact length of this wall is to multiply the number of bricks in the wall by the basic unit of dimension, then subtract one mortar joint:

7 bricks x 8" = 56"	7 bricks x 203 mm = 1421 mm
56" - 3/8" joint = 55 5/8"	1421 mm - 9 mm joint = 1412 mm
55 5/8" = 4'-7 5/8"	

A wall that stretches from an outside corner to an inside corner, such as the one in the center of Figure 8.22, has the same number of mortar joints as it does bricks. Thus this 6-brick-long wall is simply 48" (1218 mm) long.

Openings in masonry walls have one more mortar joint than masonry units. The opening to the right in Figure 8.22 is therefore figured as:

4 bricks x 8" = 32"	4 bricks x 203 mm = 812 mm;
32" + 3/8" joint = 32 3/8"	812 mm + 9 mm joint = 821 mm
32 3/8" = 2'-8 3/8"	

A wall that stretches from one inside corner to another inside corner also has one more mortar joint than masonry units.

Many types of bricks, though not all, are proportioned so that one brick length plus a mortar joint is equal to two brick widths plus two

(continued from previous page)
mortar joints. Thus, modular brickwork is dimensioned in length increments of 4" (101.5 mm). (A millimeter is very small compared to the thickness of a mortar joint, and the head joints in brickwork can be squeezed enough to allow dimensioning to a more convenient module of 100 mm).

Height dimensions in masonry are figured in a similar way. The majority of bricks are dimensioned so that three courses of brick plus three bed joints of mortar add up to 8" (200 mm). This is convenient because three courses of brick exactly match the height of one course of concrete blocks. For bricks and concrete masonry units that are made to other dimensions than these, different basic units of length and height must be computed, but the general principles are the same.

1. The small retail building whose plan is drawn below is to be built of modular bricks. Before construction can begin, you must work out exact dimensions to guide masons. Count squares to determine each dimension approximately, then fill in the exact dimensions of the brickwork, accurate to the nearest 1/8" or 1 mm, in such a way that only full bricks and half bricks need be used in the stretcher courses. Check your work by adding each chain of short dimensions and comparing the sum to the corresponding overall dimension.

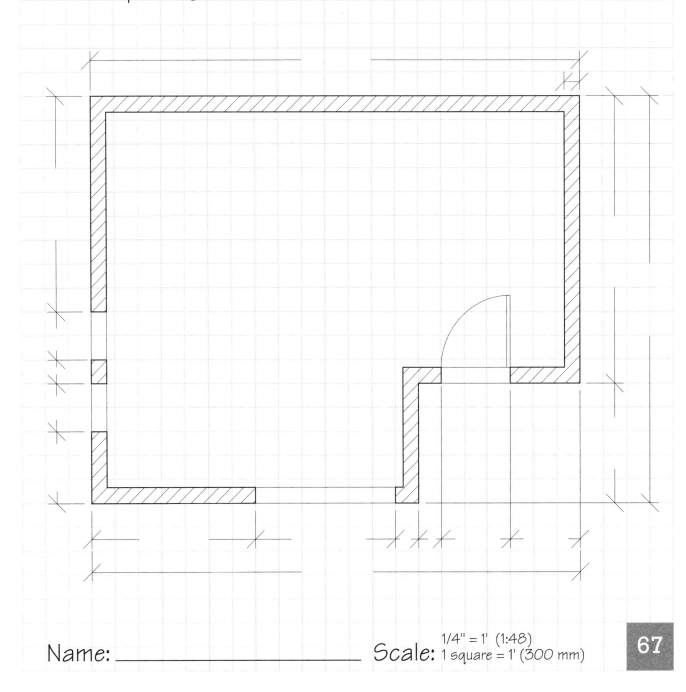

Name: _____ Scale: 1/4" = 1' (1:48)
1 square = 1' (300 mm)

2. The ceiling of this building will be flat and constructed of wood joists. If a ceiling height of approximately 9'-6" (2896 mm) is desired, figure the number of courses and the exact height of the wall for each of the following types of masonry units:

	Number of Courses	Exact Height
a. Modular brick		
b. Engineer Standard brick		
c. Closure Standard brick		
d. Roman brick		
e. Norman brick		
f. King Size brick, 2 5/8" high		
g. 8" x 8" x 16" concrete block (194 x 194 x 397 mm)		
h. Arizona adobe brick (4" x 12" x 8" with 1/2" joints) (102 x 305 x 203 mm with 13 mm joints)		

1. Draw in both elevation and section an appropriate design to span each of these openings:

   a. Doorway in a garden wall of Flemish Bond modular brickwork two wythes thick. Draw yourself to scale, standing in the opening, before you start designing the opening. You may use special brick shapes if you wish.

Top of wall ───────

Width of Doorway

Grade ───

Elevation

Section

Name: _____

Scale: 3/4" = 1' (1:16)
1 square = 4" (100 mm)

2. A window opening in a downtown apartment building built of Closure Standard bricks is two wythes thick. Draw yourself to scale in the window, and pay attention to how you detail the brickwork at the sill and jambs. Use any bond you wish, and special brick shapes as you see fit.

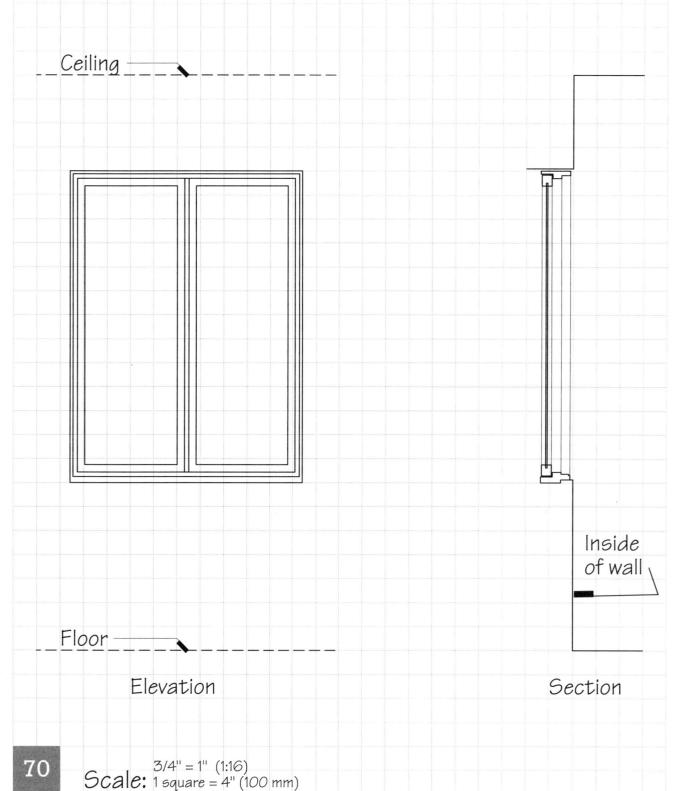

Ceiling

Inside of wall

Floor

Elevation

Section

Scale: 3/4" = 1" (1:16)
1 square = 4" (100 mm)

9

STONE AND CONCRETE MASONRY

In this exercise you will select concrete masonry units and mortar suitable to various project conditions, and detail a section through a composite masonry wall consisting of both common brick and CMU. Refer to figures 8.5, 9.21 through 9.23 of the text for information helpful to completing these tasks. The following are some additional selection guidelines.

Medium weight concrete masonry units are typically the most economically produced and the most frequently specified. Lightweight or normal weight units are used only where other considerations outway the higher production costs associated with these units. For example, lightweight units may be specified for a higher fire-resistance rating (the lighter units conduct heat more slowly), lower shipping costs, or reduced labor costs (lighter units are easier to handle). Units in the higher range of the medium weight classification, or even heavier normal weight units, may at times be specified for their higher compressive strength, lower water absorption, greater durability, increased resistance to sound transmission, or increased thermal mass.

For loadbearing concrete masonry, 8-inch (200 mm) wide units are most common, though 6-inch (150 mm) wide units may be used where a more narrow width is desired. Four-inch (100 mm) wide units are generally reserved for nonbearing applications, as part of multi-wythe composite wall construction, or for the veneer wythe of a cavity wall. Note that the core openings in 4-inch (100 mm) cmu are too narrow to easily add significant vertical reinforcing. Wider units may be used where a thicker wall or greater bearing capacity is required. For example, a composite wall consisting of an 8-inch (200 mm) wide CMU and 4-inch (100 mm) brick might be suppoted on a foundation wall constructed from 12-inch (300 mm) wide concrete masonry units.

1. Specify hollow concrete masonry units and mortar for each of the following situations:

Concrete Masonry Units		Mortar Type
Thickness & Shape	Weight	

a. Interior nonbearing partitions in a 15-story building

b. Exterior bearing walls of a large warehouse in Tampa, Florida.

c. Reinforced headers over window openings in the warehouse above

d. Heavily reinforced exterior loadbearing walls of a 17-story building in Cleveland, Ohio

e. Backup wythe of a vertically reinforced cavity wall, working in constrained-height conditions

f. CMU face veneer for the cavity wall described above

g. Basement wall for a small single-family residence, to be parged and dampproofed.

Name: _____

2. The header unit shown in Figure 9.22 of the text (the block in the second row from the top, third from the left) is formed so that the tails of a header course in a modular brick facing wythe, together with surrounding mortar, will sit neatly in the notch in the block. Centered on the wall footing below, draw a section of a composite wall consisting of an 8" (200 mm) concrete block backup and an exterior facing wythe of Common Bond modular brickwork, using header blocks as necessary. (There is no cavity between the wythes in this type of wall, only a collar joint of mortar.) Draw the thickness of the mortar joints accurately, and use a weathered joint.

Exterior

Interior

74

Scale: 1 1/2" = 1' (1:8)
1 square = 2" (50 mm)

1. Design a ceremonial gateway or archway of cut limestone for a college campus. The opening should be about 12' (3660 mm) wide. Your design may be simple or elaborate; scan pages 304 through 320 of the text for ideas. Draw the principal elevation of the gateway on this page. Indicate all the joints in the stonework.

2. Based on close examination of Figures 9.13 and 9.14 of the text, draw a horizontal section and a vertical section showing how the masonry of your gateway design is assembled. Indicate the scale you have used.

Scale:

10

MASONRY LOADBEARING WALL CONSTRUCTION

Movement Joints in Masonry Construction

For this exercise, review pages 349 through 352 in the text and the examples below.

Problem: A masonry cavity wall with brick veneer has vertical expansion joints spaced at 24' (7315 mm) o.c. Assuming a maximum change in brick temperature from cold winter nights to sunny summer days of 120°F (67°C), what is the maximum movement that will occur across each joint?

Solution: The change in length of the wall is calculated as:

$$\Delta L = \varepsilon L \Delta T$$

where

ΔL = change in length
ε = coefficient of thermal expansion
L = total length
ΔT = change in temperature

Figure 10.14 of the text gives the coefficient of thermal expansion for brick as 0.0000036 in/in/°F. Substituting into the equation:

$$\Delta L = (3.6 \times 10^{-6} \text{in/in/}°F)(24' \times 12''/1')(120°F)$$
$$\Delta L = .124'', \text{ or } 1/8'' \text{ (3 mm)}$$

Problem: What is the required joint width for the expansion joint described above, assuming that the sealant used to seal the joint has a movement capacity of 25%?

Solution: The required joint width is determined as:

$$W = \frac{100}{X} \Delta L$$

where

W = joint width
X = sealant movement capacity, percent

Substituting,

$$W = \frac{100}{25}(1/8'') = 1/2'' \text{ (13 mm)}$$

1. A long brick wall has a vertical expansion joint every 125' (38 m) of length. Assuming a maximum change in masonry temperature of 140 °F (78°C), what is the expected joint movement? Show all calculations.

2. For the movement calculated above, what is the required joint width, assuming a sealant with movement capacity of 25%? What is the required width for a sealant with movement capacity of 50%? Show calculations.

3. In the space below, draw a plan detail of the wall and one of the expansion joints you have calculated above. Use an overall wall construction like that shown in Figure 10.4 of the text. Review Figure 10.18 of the text for ideas on how to construct the joint itself. Label all joint components.

Centerline of joint

Inside face of wall

Outside face of wall

Name: _____

4. Figure 10.4, detail C shows a soft joint installed below a steel shelf angle to accommodate movement between the brick veneer and the supporting structure. Show calculations:

a. The shelf angles are spaced 12' (3660 mm) o.c. vertically. What is the expected joint movement due to brick thermal expansion, assuming a change in temperature of 120°F (67°C)?

_____

b. Assuming a coefficient of moisture expansion of the brick (due to gradual absorption of moisture by the kiln-dried brick) of 0.0002 in/in (mm/mm), what is the expected moisture-induced increase in height of a 12' (3660 mm) section of the brick veneer?

_____

c. Assuming a coefficient of drying shrinkage of the concrete structure of 0.0005 in/in (mm/mm), and an additional 0.001 in/in (mm/mm) for structural creep, what is the expected decrease in height of 12' (3660 mm) of the concrete structure?

_____

d. What is the total expected joint movement, due to all of the above? (Add together all of the above.)

_____

e. Assuming a joint sealant with movement capacity of 50%, what is the required joint width?

_____

f. Add 1/8" (3 mm) to the calculated joint width to allow for variations in construction tolerance. What is the final design joint width?

_____

1. On the concrete slab edge shown below, draw a section through the base of an unreinforced wall consisting of an outer wythe of Standard bricks, a 2" (51 mm) uninsulated cavity, and an inner wythe of 8" (200 mm) hollow concrete masonry units. Show and label every necessary feature of the construction. See Figures 10.1, 10.4, and 10.10 in the text for guidance.

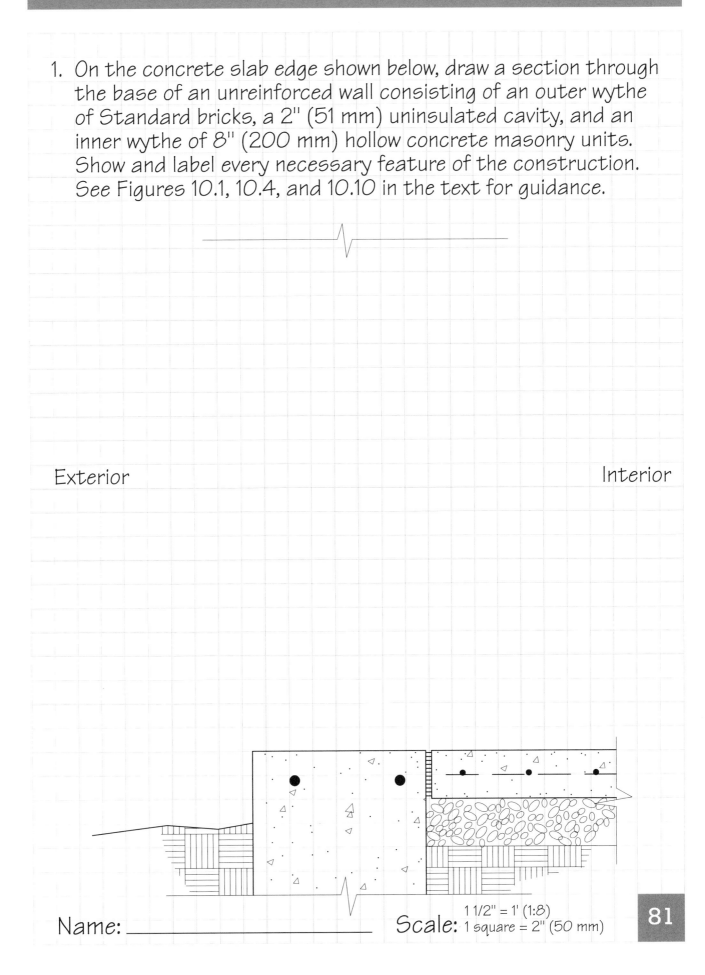

Exterior

Interior

Name: _____

Scale: 1 1/2" = 1' (1:8)
1 square = 2" (50 mm)

2. Draw in section and label completely the base of a cavity wall consisting of a full 3" (76 mm) outer wythe of ashlar limestone blocks, an inner wythe of 8" (200 mm) hollow concrete masonry units, and an inch (25.4 mm) of insulating foam plus gypsum board as shown in Figure 23.5 of the text. Make the stone blocks nominally 18" (457 mm) high, and mount them in the manner shown in Figure 9.11 of the text.

Exterior   Interior

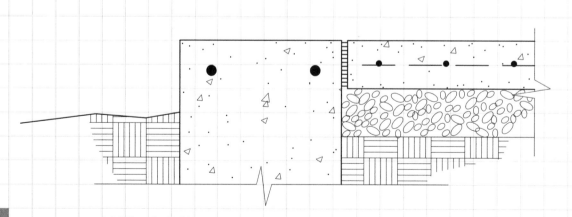

Scale: 1 1/2" = 1' (1:8)
        1 square = 2" (50 mm)

11

STEEL FRAME CONSTRUCTION

This exercise will help you become more familiar with the dimensional properties of steel structural shapes. You will need to refer to Figures 11.12 and 11.13 of the text. Keep in mind that you are looking here at only a small sampling of the available sizes of steel members. Wide-flange shapes, for example, are available in depths ranging from 4" to 36" (100 to 900 mm), although the relative proportions of the shapes are more or less constant regardless of size.

Figure 11.12 is divided vertically into two parts. The second part gives structural properties of the shapes that will be familiar to those of you who have studied structural engineering. All the information you need to complete this exercise is contained in the first part. You will see that the cross-sectional area of each shape is given, along with its actual depth and the detailed dimensions of its flanges and web. The distances T, k, and k_1 are particularly useful; they locate the point at which the curved fillet begins between the flange and the web. T is also the maximum length of plate or angle that can be fastened to the web.

Notice that each grouping of shapes in the table shares the same nominal flange width and nominal overall depth. T is constant within each grouping because the same interior roller is used throughout, with only the outside rollers being moved to create the different weights (see Figure 11.10 in the text). The shapes with 12" and 10" flange widths are used largely for columns and H-piles, while those with the narrower flanges are used primarily for beams.

1. Draw accurate sections of the shapes indicated, each centered on the given centerline with its lower edge on the solid line.

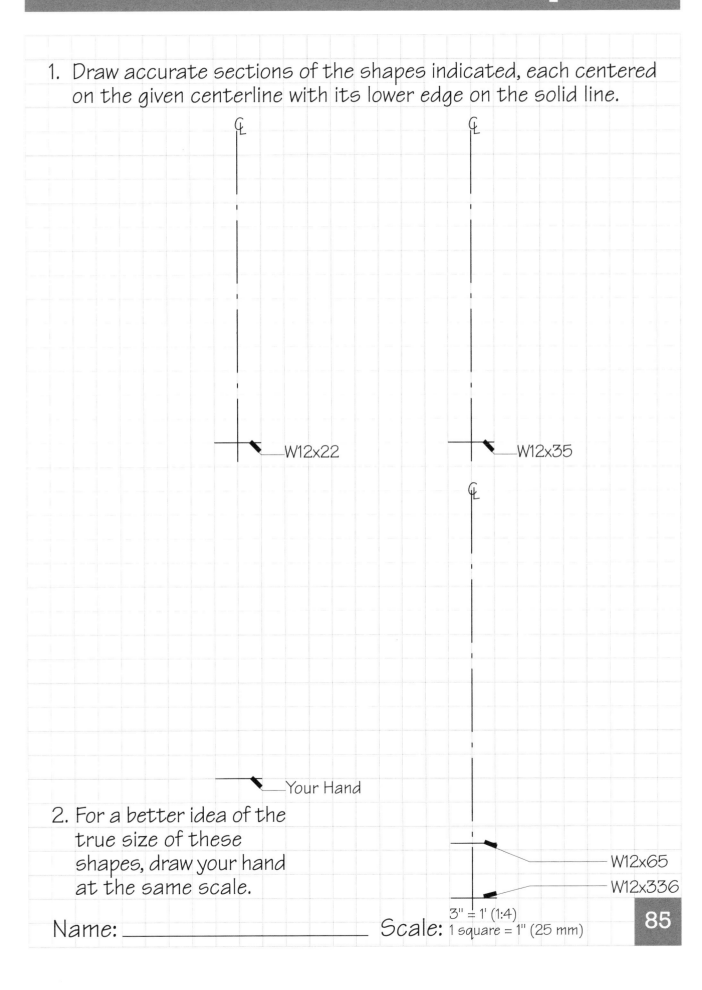

℄

℄

W12x22

W12x35

℄

Your Hand

2. For a better idea of the true size of these shapes, draw your hand at the same scale.

W12x65

W12x336

Name: _____   Scale: 3" = 1' (1:4)
1 square = 1" (25 mm)

3. Draw accurate cross sections of the angle shapes indicated.

L4x3x1/4
L4x3x5/8

L3x3x3/16
L3x3x1/2

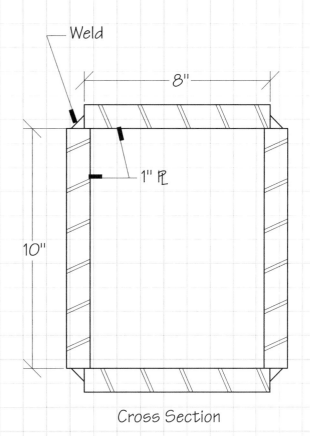

Weld

8"

1" PL

10"

Cross Section

4. How much will a one-foot (305 mm) length of this built-up steel box column weigh, including the corner welds? Hint: Calculate the cross-sectional area of the steel in this column, and relate it to an area/weight ratio calculated from Figure 11.12 of the text, or use the density of steel from Figure 11.91. Show calculations.

Scale: 3" = 1' (1:4)
1 square = 1" (25 mm)

The complete structural design of a steel building frame is an involved process, but it begins with the laying out of a framing plan, which can be rather simple for many buildings. See Figure 11.42 in the text for example of a typical structural steel framing plan.

Usually the bay spacings in a steel frame are kept to about 36' (11 m) or less in order to minimize the size of the beams and girders, and bay sizes are kept constant except where interruptions such as elevator shafts and stairs occur. A good way to begin laying out a framing plan is to use freehand overlays on tracing paper to try dividing the building plan into a number of different sizes and shapes of bays, until one layout shows promise of working better than the others.

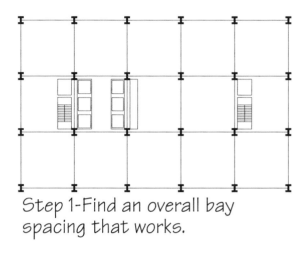

Step 1-Find an overall bay spacing that works.

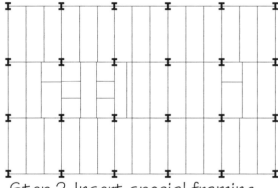

Step 2-Insert special framing around openings, then add the rest of the framing.

Then special arrangements of beams and girders, much like the headers and trimmers used around openings in platform frame wood floors, must be designed to frame around stairs and shafts. Move lines of columns on tracing paper until you arrive at a simple, logical layout that avoids excessive irregularities. (Usually the architectural plan can be adjusted slightly if necessary to arrive at a satisfactory framing plan.) Check to be sure that the layout does not involve excessively long spans, which are costly, or spans that are so short that they require too many columns and/or cut the habitable space of the building into too many little pieces.

Within a typical bay, the layout of girders, beams, and decking should be done with the aid of the rules of thumb on page 373 of

(continued from previous page)
the text. Select a trial depth and type of deck, and lay out girders to support the beams or joists. Determine preliminary depths for each of these members--are they reasonable? If not, adjust spacings and sizes until they are.

You are designing an 8- to 10-story regional office building in downtown Omaha, Nebraska, for Associated Mutual Casualty and Life Corporation. Three possible plan arrangements for a typical floor of the building are shown below and on the following page. Draw a feasible framing plan over each of the plans, and give approximate depths for the typical beams and girders. Assume that you will use W10 columns.

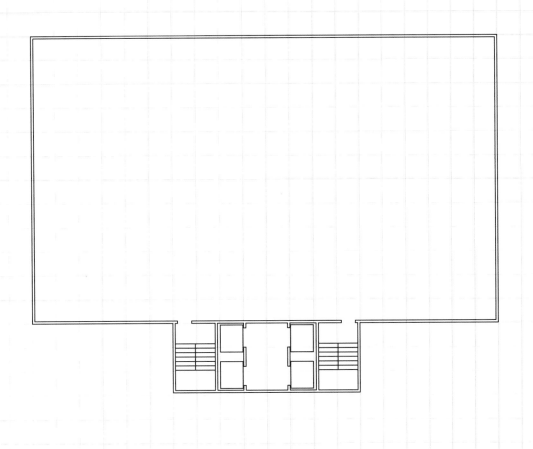

Name: _____ Scale: 1" = 20' (1:250)
1 square = 5'-0" (1.5 m)

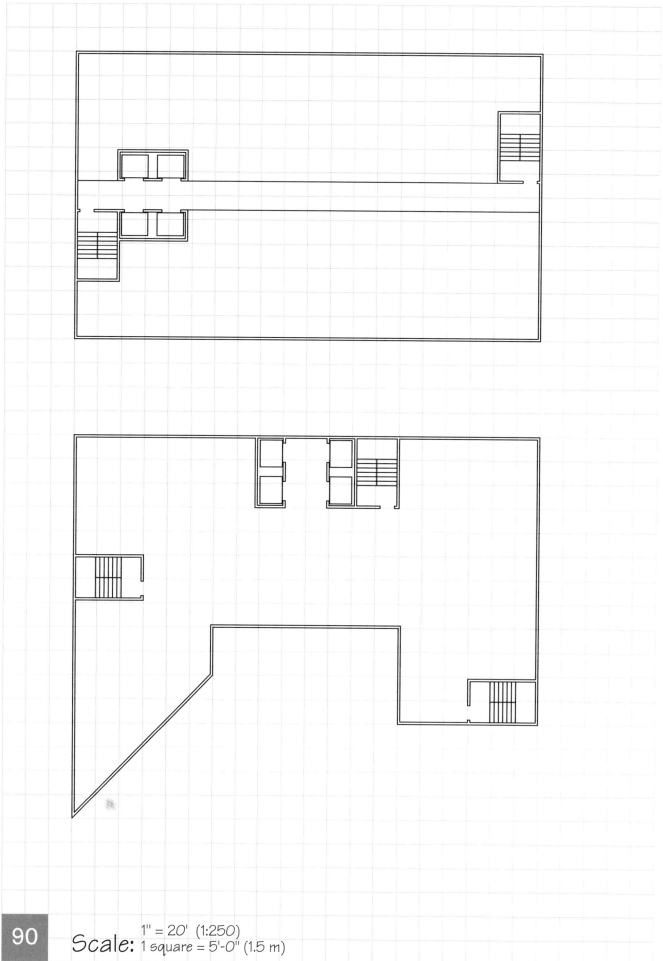

Scale: 1" = 20' (1:250)
1 square = 5'-0" (1.5 m)

Structural steel connections are designed on the basis of the loads they must transmit from one member to another, which is beyond the scope of this course, but this exercise will help you to become familiar with the more common methods of joining steel members in a building frame.

You should keep your text close by as you work on this exercise, ready to consult pages 387 through 395 as needed. Before you begin work, also familiarize yourself with the detailed information given in the tables of the following page of this workbook.

(continued from previous page)

Some Typical Framed Connections

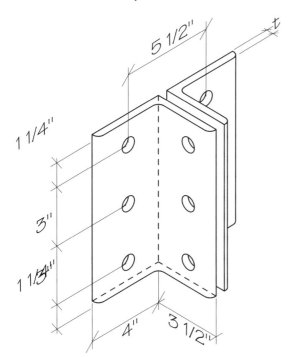

Rows of Bolts	Applicability	Length of Angle	Thickness of Angle (t)		
			3/4" bolts	7/8" bolts	1" bolts
2	W12,10,8	5 1/2'	5/16"	3/8"	7/16"
3	W18,16,14,12	8 1/2"	5/16"	3/8"	7/16"
4	W24,21,18,16	11 1/2"	5/16"	3/8"	7/16"
5	W30,27,24, 21,18	14 1/2"	5/16"	3/8"	7/16"

Some Typical Seated Connections

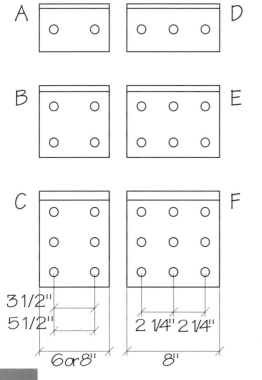

Type	Angle Size	Thickness
A, D	4x3	3/8"-5/8"
	4x3 1/2	3/8"-5/8"
	4x4	3/8"-3/4"
B, E	6x4	3/8"-7/8"
	7x4	3/8"-7/8"
	8x4	1/2"-1"
C, F	8x4	1/2"-1"
	9x4	1/2"-1"

1. Select any exterior column location from one of the framing plans you developed in Exercise 11.2 and draw details for it as specified. A W10x49 column is assumed.

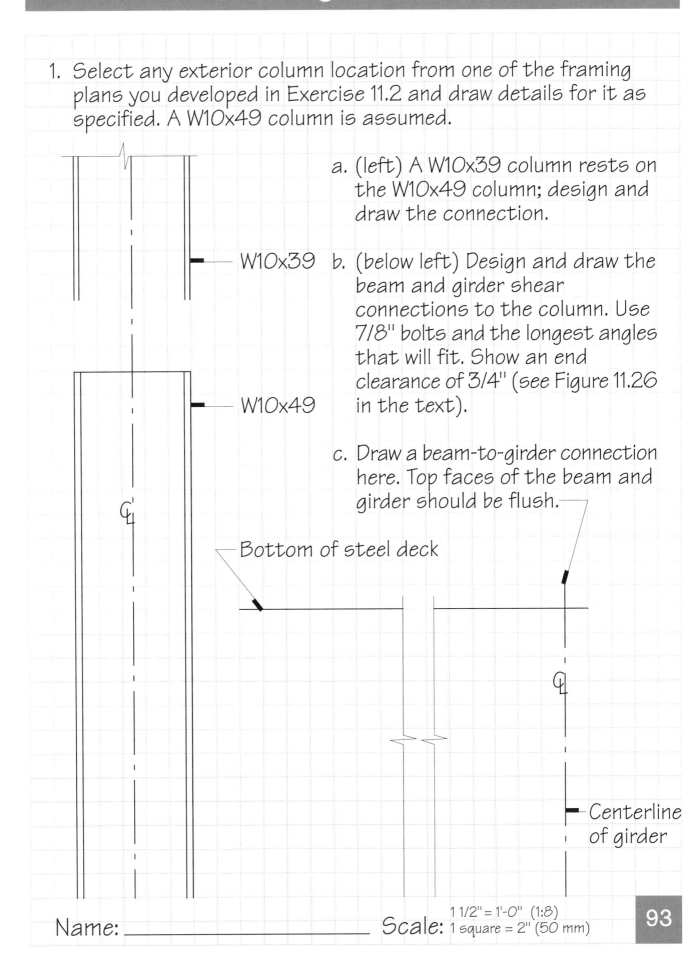

W10x39

W10x49

a. (left) A W10x39 column rests on the W10x49 column; design and draw the connection.

b. (below left) Design and draw the beam and girder shear connections to the column. Use 7/8" bolts and the longest angles that will fit. Show an end clearance of 3/4" (see Figure 11.26 in the text).

c. Draw a beam-to-girder connection here. Top faces of the beam and girder should be flush.

Bottom of steel deck

C
L

C
L

Centerline of girder

Name: _____

Scale: 1 1/2" = 1'-0" (1:8)
1 square = 2" (50 mm)

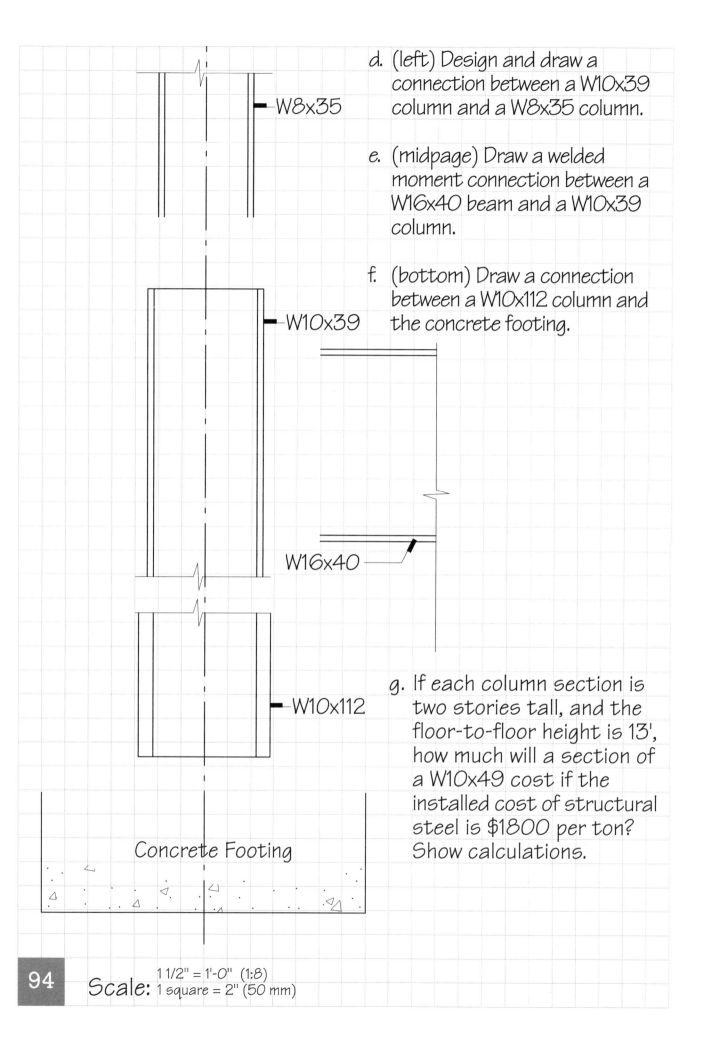

d. (left) Design and draw a connection between a W10x39 column and a W8x35 column.

W8x35

e. (midpage) Draw a welded moment connection between a W16x40 beam and a W10x39 column.

f. (bottom) Draw a connection between a W10x112 column and the concrete footing.

W10x39

W16x40

g. If each column section is two stories tall, and the floor-to-floor height is 13', how much will a section of a W10x49 cost if the installed cost of structural steel is $1800 per ton? Show calculations.

W10x112

Concrete Footing

Scale: 1 1/2" = 1'-0" (1:8)
1 square = 2" (50 mm)

The city of Des Moines, Iowa has contracted with you for the design of a covered marketplace. This will be a roofed space without walls, 100' x 120' (30 x 37 m) in plan, in which stalls will be set up on weekends to sell fresh vegetables and fruits, eggs, crafts, and antiques. The city asks that your design be light and airy, and that it have daylighting throughout. The city engineer suggests that to save construction time, your design should use only wide-flange shapes, open-web steel joists, and steel decking, which require little fabrication.

1. Draw a freehand perspective showing the character of the interior of the marketplace you have designed.

2. Draw a framing plan of your marketplace design, giving approximate sizes of all members, in the space indicated below.

3. On the ground line below, draw either a cross section or a longitudinal section to show the shape of the roof and to indicate how it is braced against wind loads.

Grade

96

Scale: 1" = 20' (1:240)
1 square = 5" (1.5 m)

4. On this page and the following page, draw the principal connections for the frame you have designed. If your design relies on AISC Type I connections for lateral stability, this should be reflected in these details.

Scale: 1 1/2" = 1' (1:8)
1 square = 2" (50 mm)

12

LIGHT GAUGE STEEL FRAMING

12.1 Light Gauge Steel
    Framing Details

Light Gauge Steel Framing Details

Light gauge steel framing is similar to wood light framing, both in the standardized sizes of the framing members, and the manner in which the various joists, studs, and rafters are assembled into framed structures. Both systems also share many of the same interior and exterior finish systems.

For guidance with the following exercise, refer to Figures 12.1 through 12.5 of the text. You may also want to look back at Exercise 6.1 in this workbook to compare the light gauge steel framing details you develop here with their wood light frame counterparts.

1. Complete all light gauge steel components in the following wall section, including floor joists, wall studs, roof rafters, and required clips, angles, stiffeners, and fasteners. Label all components.

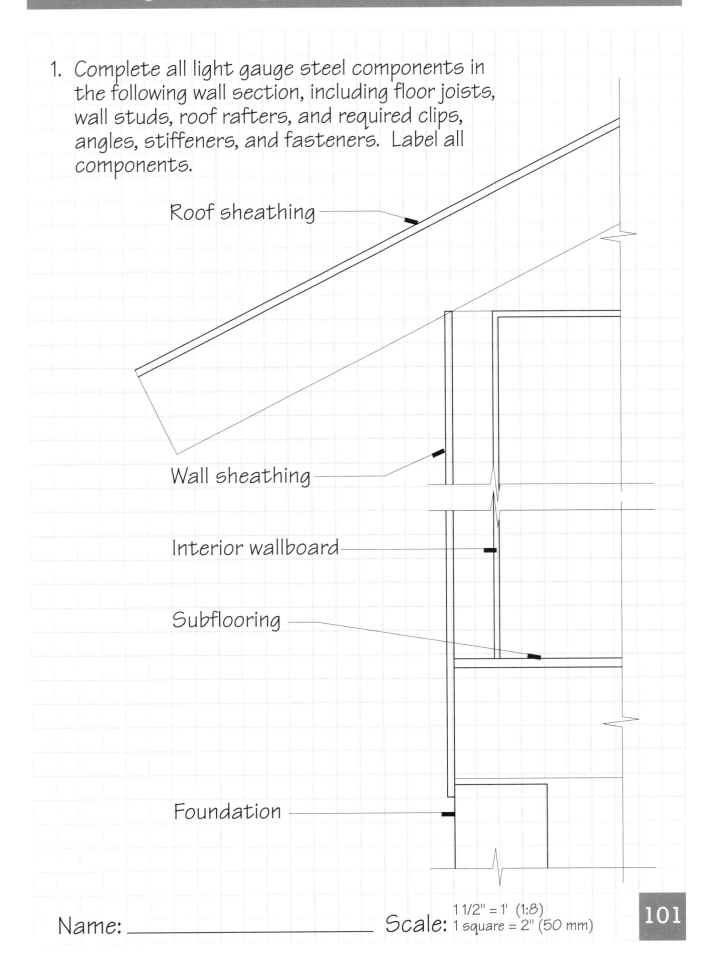

Roof sheathing

Wall sheathing

Interior wallboard

Subflooring

Foundation

Name: _____

Scale: 1 1/2" = 1' (1:8)
1 square = 2" (50 mm)

2. Using the same size framing members as in part 1, complete the light gauge steel framing for the following rake, door head, and exterior wall corner plan details. Show all accessories, and label all components.

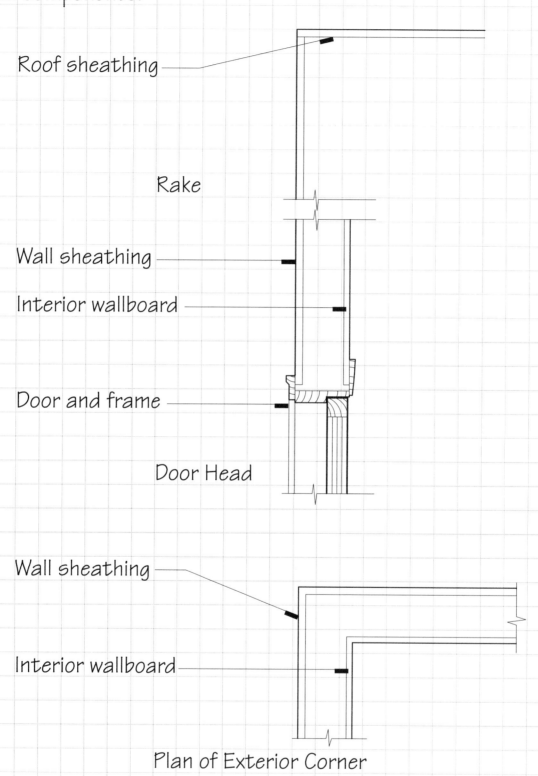

Roof sheathing

Rake

Wall sheathing

Interior wallboard

Door and frame

Door Head

Wall sheathing

Interior wallboard

Plan of Exterior Corner

Scale: 1 1/2" = 1' (1:8)
1 square = 2" (50 mm)

CONCRETE CONSTRUCTION

13.1 Detailing Concrete
     Reinforcing Bars

Reinforcing bars are bent and placed in accordance with ACI 318, which is a comprehensive standard governing every aspect of reinforced concrete construction. This exercise will familiarize you with some of its provisions that relate to detailing of concrete reinforcing bars. To complete this exercise successfully you will need to refer first to Figures 13.19, 13.23, 13.25 and 13.26 of the text.

When referring to Figure 13.23, which is based on ACI 318, note the following: "D" and "d" have different meanings. Uppercase "D" refers to the diameter of a hook bend. Lowercase "d" refers to the nominal diameter of the bar itself. Also, where the following exercise asks for critical dimensions of hooks, you should ignore dimenions labeled as "A or G" or "J". These dimensions are used by the fabricator when cutting bars, but they do not correspond precisely to the minimum dimensions required by the standard.

ACI 318 also specifies other critical dimensions related to reinforcing steel placement. Primary reinforcing and stirrups in beams must lie beneath a concrete cover of at least 1 1/2" (38 mm). Parallel reinforcing bars must be separated by a clear distance of one inch (25.4 mm) or the diameter of the bars, whichever is greater. At the end of a concrete beam, the bend that anchors the top bars should reach a point two inches (51 mm) from the outside face of the column, and the bottom bars must extend six inches (152 mm) into the column.

The first stirrup in a beam is usually placed at a distance of 2" (51 mm) from the inside face of the column, and spacings thereafter are determined by the engineer or architect who designs the beam.

There are also standards in ACI 318 that deal with bar placements at interior columns, and in slabs, walls, footings, and other types of concrete structures.

Shown below is a larger-scale elevation view of the left end of the continuous concrete beam shown in Figure 13.28 of the text. The structural plans call for a 10" wide by 18" deep (254 by 460 mm) beam with two #9 top bars and two #8 bottom bars. Fifteen #3 U-stirrups are required at a spacing of 7" (178 mm).

1. Make an accurately scaled drawing of the reinforcing bars in the beam, including top bars, bottom bars, and stirrups, drawing them as if the concrete were transparent. Give critical dimensions of any hooks. Do not show column reinforcing or bar deformations.

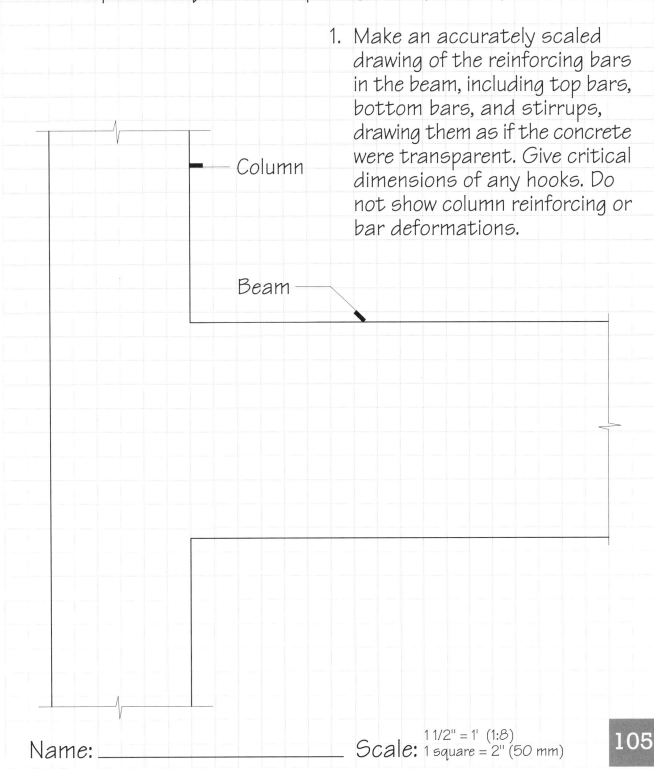

Column

Beam

Name: _____

Scale: 1 1/2" = 1' (1:8)
1 square = 2" (50 mm)

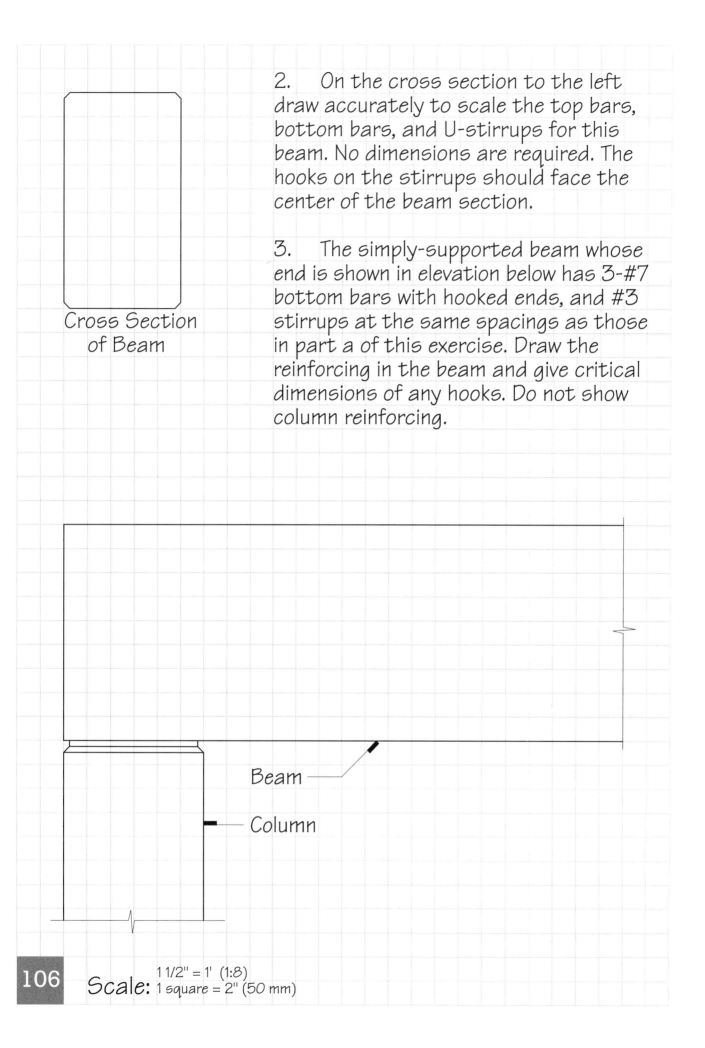

Cross Section
of Beam

2. On the cross section to the left draw accurately to scale the top bars, bottom bars, and U-stirrups for this beam. No dimensions are required. The hooks on the stirrups should face the center of the beam section.

3. The simply-supported beam whose end is shown in elevation below has 3-#7 bottom bars with hooked ends, and #3 stirrups at the same spacings as those in part a of this exercise. Draw the reinforcing in the beam and give critical dimensions of any hooks. Do not show column reinforcing.

Beam

Column

Scale: 1 1/2" = 1' (1:8)
1 square = 2" (50 mm)

14

SITECAST CONCRETE FRAMING SYSTEMS

Sitecast Concrete Framing Systems

There are three primary technical questions to consider when selecting a sitecast concrete framing system:

1. Are the structural bays square in proportion, or are they rectangular? If the bays are square or nearly so, a two-way system is preferred because it is more economical of steel and concrete; for rectangular bays, one-way systems must be used.

2. How long are the spans? For spans up to approximately 35' (11 m), solid slab or plate systems are appropriate, but longer spans usually require a joist system.

3. How heavy are the floor or roof loadings? Heavier loadings require deeper beams and girders, and thicker slabs. Systems with deep column to slab connections, such as the flat slab, also have superior load carrying capacity in comparison to systems with thinner connections, such as the flat plate. (Residences, hotels, offices, and classrooms have light floor loadings while industrial, storage and auditorium buildings have heavier loadings.)

See page 533 in the text for more factors to consider in the selection of sitecast concrete framing systems.

To solve the problems in this first exercise, you will need to consult a number of the illustrations in Chapter 14 of the text, as well as the preliminary design guidelines on page 536, which will help you determine the approximate size and thicknesses for the various components of any system. If you are adopting a joist system, use these values also to determine the depth of pan or dome, then refer to figures earlier in the chapter to learn typical dimensions of the system in detail.

1. For each building in this exercise, the column locations for a typical bay are shown to the right. Select and name an appropriate, economical concrete framing system for each. Complete the framing plan by showing girders, beams, joists, or drop panels carefully drawn to scale, and by indicating the approximate dimensions of each feature. Draw a typical section to the left, at the scale indicated, showing details of the concrete elements and typical locations of reinforcing bars. Extend each plan or section drawing to the limits of the box in which it is drawn.

a. Metal casting plant (heavy floor loading)	System selected:
Typical Section	Framing Plan

b. Hotel (light floor loading)	System selected:
Typical Section	Framing Plan

Scale: 3/4" = 1' (1:16) 1 square = 4" (100 mm)

Scale: 1/16" = 1' (1:192) 1 square = 4' (1.2 m)

Name: _____

c. Elementary School (light floor loading)	System selected:
Typical Section	Framing Plan

d. Office building (light floor loading)	System selected:
Typical Section	Framing Plan

e. Paper warehouse (heavy floor loading)	System selected:
Typical Section	Framing Plan

Scale: 3/4" = 1' (1:16)
1 square = 4" (100 mm)

Scale: 1/16" = 1' (1:192)
1 square = 4' (1.2 m)

1. Design a ceremonial gateway or archway of sitecast concrete for a college campus. The opening should be about 12' (3660 mm) wide. Your design may be simple or elaborate. Scan pages 539 to 542 of the text for ideas, with special reference to Figures 14.51 and 14.52. Draw the principal elevation of the gateway on this page. Indicate surface finishes or textures, form tie holes, recess strips, and other features. Indicate at the lower right the scale you have used.

2. Draw a horizontal section and a vertical section showing the concrete shape and arrangement of reinforcing of the primary parts of your gateway design. Reinforcing sizes and dimensions are not required.

Scale:

15

PRECAST CONCRETE FRAMING SYSTEMS

Precast Concrete Framing Systems

Most precast, prestressed concrete buildings can be framed with very simple layouts of standard slab and beam elements. Before designing framing systems for the buildings whose plans are shown in Exercise 15.1, you will want to review pages 560 through 564 of the text. Take special note of the Preliminary Design guidelines provided on page 563 and Figures 15.5 through 15.7.

To select a precast, prestressed slab element that is appropriate for a given building design, first examine the rough floor plan to locate lines of support--either beams or loadbearing walls. Then determine the maximum span between lines of support, and consult the Preliminary Design guidelines referenced above to find the type or types of slab element that can do the job. Make your selection and draw the joints between the slabs on the plan, adding beams or lintels as necessary.

Preliminary floor plans for five different buildings are shown this page and the next. For each, determine a type and approximate depth of precast, prestressed slab element for the floor or roof. Draw the slab elements diagrammatically on each plan, along with any necessary beams or lintels. The last floor plan represents a more complicated problem than the others, and requires that you make some design judgements concerning column locations and the character of the exterior walls of the building.

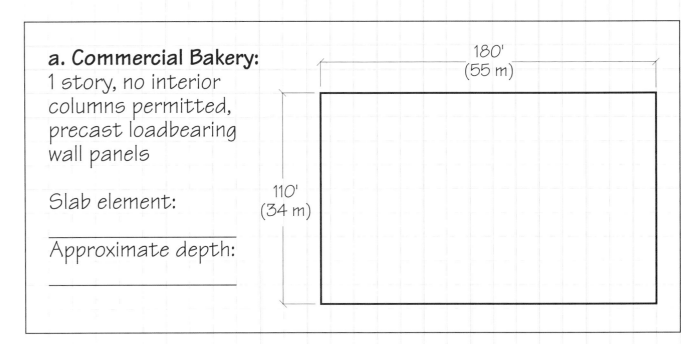

a. Commercial Bakery:
1 story, no interior columns permitted, precast loadbearing wall panels

Slab element:
_____

Approximate depth:
_____

180'
(55 m)

110'
(34 m)

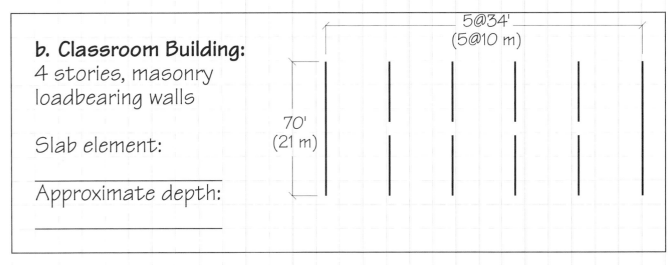

b. Classroom Building:
4 stories, masonry loadbearing walls

Slab element:
_____

Approximate depth:
_____

5@34'
(5@10 m)

70'
(21 m)

Name: _____

Scale: 1" = 50' (1:600)
1 square = 12'-6" (3.8 m)

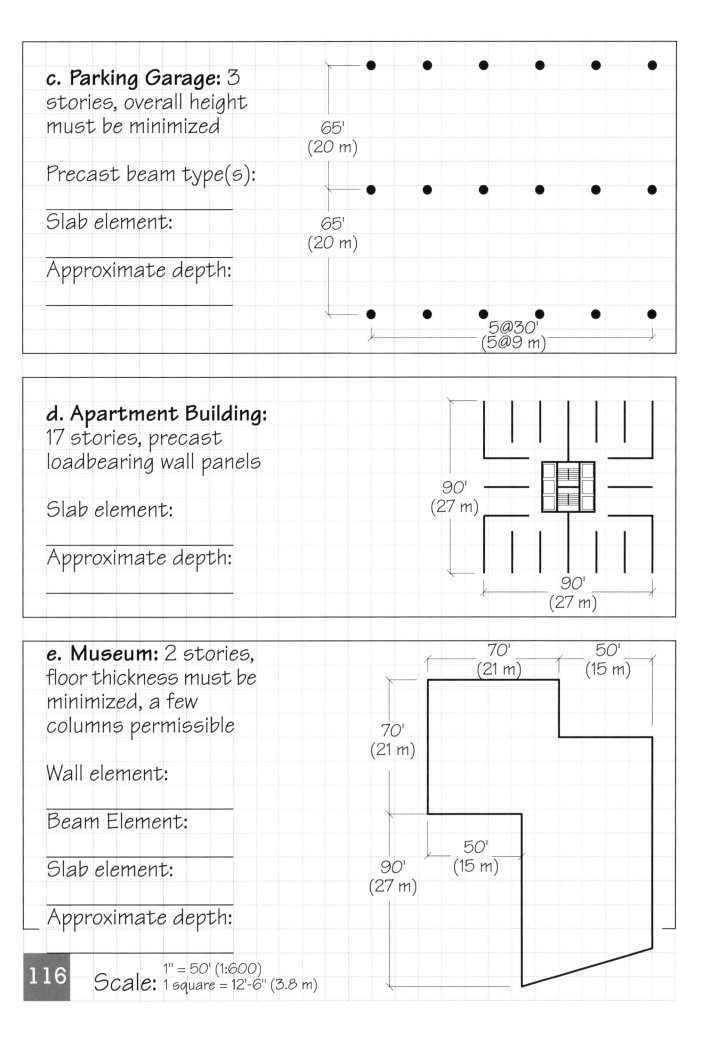

c. Parking Garage: 3 stories, overall height must be minimized

Precast beam type(s):

_____

Slab element:

_____

Approximate depth:

_____

65'
(20 m)

65'
(20 m)

5@30'
(5@9 m)

d. Apartment Building: 17 stories, precast loadbearing wall panels

Slab element:

_____

Approximate depth:

_____

90'
(27 m)

90'
(27 m)

e. Museum: 2 stories, floor thickness must be minimized, a few columns permissible

Wall element:

_____

Beam Element:

_____

Slab element:

_____

Approximate depth:

_____

70'
(21 m)

50'
(15 m)

70'
(21 m)

90'
(27 m)

50'
(15 m)

116

Scale: 1" = 50' (1:600)
1 square = 12'-6" (3.8 m)

In the exercise that follows, you will be asked to draw two large-scale details of precast concrete connections for a building whose framing plans you designed in the previous exercise. In preparation, you should review Figures 15.5-15.7, 15.13-15.22, 15.26, and 15.26 in the text. None of these details is exactly the solution needed in this exercise, but you may wish to use one or two of them as starting points for your work. A rough layout of each detail on scratch paper will allow you to work out the bugs before committing the drawing to the final sheet.

As a starting point for each detail, draw the end of the slab element to scale, making use of the detailed dimensions given on the following page. The beam size will be largely determined by the size of the slab; chances are good that detailed structural calculations would show this size to be appropriate, subject to some adjustment of the number and size of prestressing strands. The column need not be as wide as the bottom of the beam if the loads on it are not exceptionally heavy.

Remember to add necessary bearing pads, spacers, weld plates, grout, topping, and other detailed features to your drawings.

Some Typical Dimensions of Precast Concrete Elements

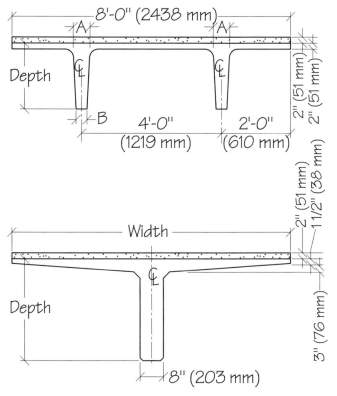

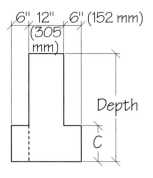

Double Tees

Depth	A	B
12"-24"	5 3/4"	3 3/4"
(300-600 mm)	(164 mm)	(95 mm)
32"	7 3/4"	4 3/4"
(800 mm)	(197 mm)	(121 mm)

Single Tees

Depth	Width
36"	8'-0"
(914 mm)	(2438 mm)
48"	10'-0"
(1219 mm)	(2540 mm)

Inverted Tee Beams and L-Shaped Beams

Depth	C
20"	8"
(508 mm)	(203 mm)
24"-36"	12"
(610-914 mm)	(305 mm)
40"-60"	16"
(1016-1524 mm)	(406 mm)

Depths increase in 4" (102 mm) increments.

Columns

12"x12"	(305x305 mm)
14"x14"	(356x356 mm)
16"x16"	(406x406 mm)
18"x18"	(457x457 mm)
20"x20"	(508x508 mm)
24"x24"	(610x610 mm)

1. Draw a detail section through the spandrel beam of the building whose framing plan you developed in part c of the previous exercise. Show and label the beam, the slab element, and all components of the connection between the two. Show the locations of the prestressing strands and reinforcing in both the slab and the beam.

2. Draw and label an exterior elevation of the beam/column connection for this same building, showing all details of how the components are joined. Use broken lines to delineate the slab elements that rest on the beam.

Scale: 1 1/2" = 1' (1:8)
1 square = 2" (50 mm)

16

ROOFING

Low-Slope Roof Drainage

Low-slope roofs should slope at least 1/4" per foot (1:48)--the minimum permitted by building codes for most roofs, and the minimum required by most membrane manufacturers in order to honor their product warranties. In practice, the least permitted slope is often specified so as to minimize the overall roof height. But steeper slopes of 3/8" or 1/2" per foot (1:32 or 1:24) can more efficiently remove water from the roof surface and are also common.

Most codes require secondary drains to serve as back up for main drains that may become clogged. Secondary drains are set 2" above the main drain, and may take the form of either through-wall scuppers at the parapet, or additional drains set close to each main drain, but with a screw-on collar to raise the drain level as required.

Use the following guidelines for laying out drainage for low-slope roofs:

1. Drains should serve an area no greater than 10,000 sf (900 m2), and no point further than 50' (15 m).
2. Drainage areas should be arranged so that water does not need to follow circuitous paths to reach a drain.
3. Assuming equal slope throughout the roof, use 45° angles to lay out drainage plane intersections.
4. Label high point elevations above the reference drain level. In the example at right, a 25' run requires a rise of 6 1/4" (25' x .25"/ft = 6.25").
5. Use small, triangular, sloped surfaces called "crickets" to divert water where it would otherwise collect or become trapped behind obstacles.

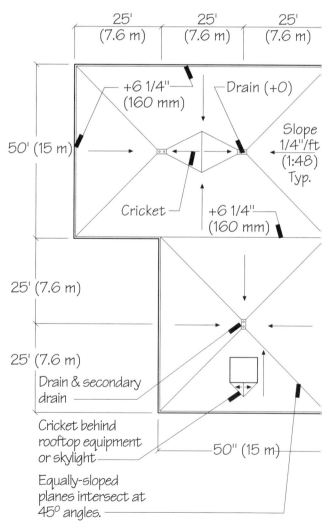

25' (7.6 m)   25' (7.6 m)   25' (7.6 m)

+6 1/4" (160 mm)   Drain (+0)

50' (15 m)

Slope 1/4"/ft (1:48) Typ.

Cricket   +6 1/4" (160 mm)

25' (7.6 m)

25' (7.6 m)

Drain & secondary drain

Cricket behind rooftop equipment or skylight

50" (15 m)

Equally-sloped planes intersect at 45° angles.

122

1. On the roof plan to the right:

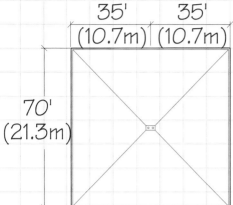

35' (10.7m) 35' (10.7m)

70' (21.3m)

   a. Add arrows indicating slope.

   b. Assuming a slope of 1/4" per foot (1:50), label roof high points.

   c. If the roof edge can rise as much as 16" (400 mm) above the drain level, what is the maximum possible slope, specified to the nearest 1/8" (1:48, 1:32, 1:24, etc.) increment? _____

2. On the roof plan below:

   a. Add arrows indicating slope.

   b. Assuming the high point on the central cricket is 6.25" (160 mm) above the drain level, what is the slope of the cricket? _____

   c. Assuming the remaining roof surfaces slope at 3/8" per foot (1:32), label the roof high points.

   d. Modify the roof plan so that the slope of the cricket matches the typical slope of the roof. (Note that in practice, either configuration may be acceptable.)

4@25' (7.6 m)

50' (15 m)

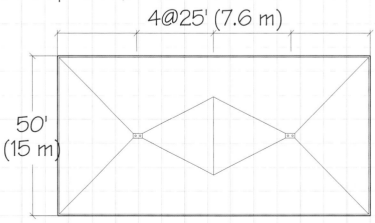

Name: _____

3. Complete the roof drainage plan below (taken from Exercise 11.2) showing roof drains, crickets, directions of slopes, magnitudes of slopes, and roof high points. Note that exterior scuppers and downspouts for collecting water from the penthouse rooftops and depositing that water onto the main roof level are already shown.

Stair and elevator penthouses. Top of penthouses 9'-0" (2.75 m) above the roof level

Scupper and downspout to main roof level.

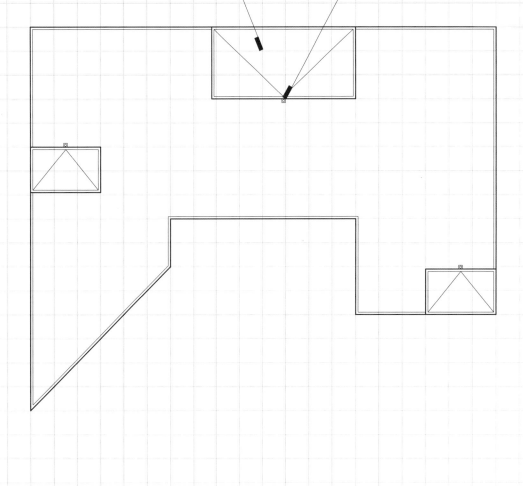

Scale: 1" = 20' (1:240)
1 square = 5'-0" (1.5 m)

In preparation for this exercise, you should review pages 609 through 621 of the text.

1. The drawing below represents a concrete masonry wall supporting a precast concrete hollow-core plank roof deck and concrete masonry parapet. Complete this detail by drawing and labeling all of the components needed to finish the roof and parapet. Use a protected membrane roof as illustrated in the upper portion of Figure 16.12 of the text, using 2" insulation boards.

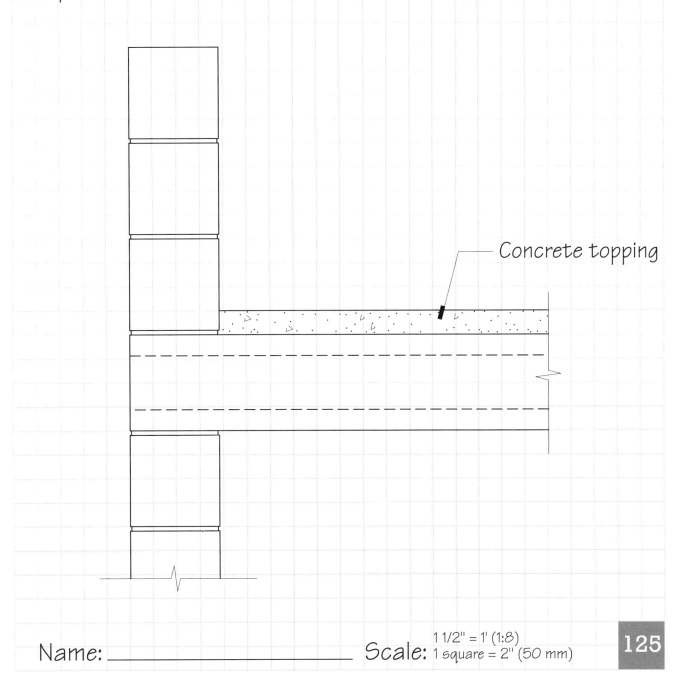

Concrete topping

Name: _____

2. A building separation joint in the roof structure of this building is shown below. Draw and label the roofing components at this joint. Use 2x12s for the curbs. Wood plates have already been attached to the concrete with powder-driven fasteners to start the detail for you.

Scale: 1 1/2" = 1' (1:8)
1 square = 2" (50 mm)

GLASS AND GLAZING

17.1 Selecting Glass
and Glazing

1. Indicate a type of glass appropriate for each of the following uses:

   a. A window in a fire door: _____
   b. Overhead sloping glazing: _____
   c. An all-glass door: _____
   d. West-facing office windows in Arizona: _____

   e. Residential windows in northern Ontario: _____

   f. A window in a public washroom: _____
   g. A north-facing clerestory: _____

2. Referring to Figures 17.14 through 17.16 of the text, complete the window sash glazing details below, drawing and labeling all glass and glazing-related components. Exterior is to the left.

Glass

Glass

a. 1/2" (13 mm) double glazing with wood stop

b. 1" (25.4 mm) double glazing with wet and/or dry glazing components

Name: _____   Scale: 1/2 Full Size (1:2)
1 square = 1/2" (13 mm)

WINDOWS AND DOORS

18.1 Selecting Windows
and Doors

1. Recommend a window type and frame material for each of the following uses.

   a. Office window in a six-story office building, no ventilation required:

   b. Classroom window in a one-story school, directly adjacent to a playground, ventilation required:

   c. Replacement window for an historic New England residence:

   d. Doors opening from a residential living space to an exterior patio, with the greatest possible openness and ventilation:

2. Recommend a door style (flush swinging, style-and-rail swinging, bifold, coiling, etc.) and material for each of the following uses:

   a. Door for office in 1.a. above, permitting partial view from corridor:

   b. Exit door for classroom for 1.b. above:

   d. Front door for the residence in 1.c. above:

   e. Door from corridor to exit stairway for office building in 1.a. above. Stairway enclosure is 2-hour rated:

Name: _____

DESIGNING CLADDING SYSTEMS

For part 1 of this exercise, review Figures 19.6 through 19.9 and pages 720 through 724 in the text to familiarize yourself with the principals of Rainscreen Design, and the components that make up this system. For part 2, review Figures 19.12 and 19.13 and pages 726 and 727 of the text for sealant joint design guidelines.

1. In the two assemblies below, identify all elements related to keeping water out of the assembly. Label them according to their rainscreen design functions including rainscreen, pressure equalization chamber, air barrier, etc.

Door
Weatherstrip
Aluminum drip

Wood Sill

Door Sill Section

Masonry facing
Cavity

Backup wall

Flashing &
weep holes

Masonry Cavity Wall Section

Name: _____   No Scale

2. Draw to scale and label each component of the sealant joints between the granite wall cladding panels shown below. Exterior is to the left. Assume that a sealant joint width of 3/8" (9.5 mm) is required.

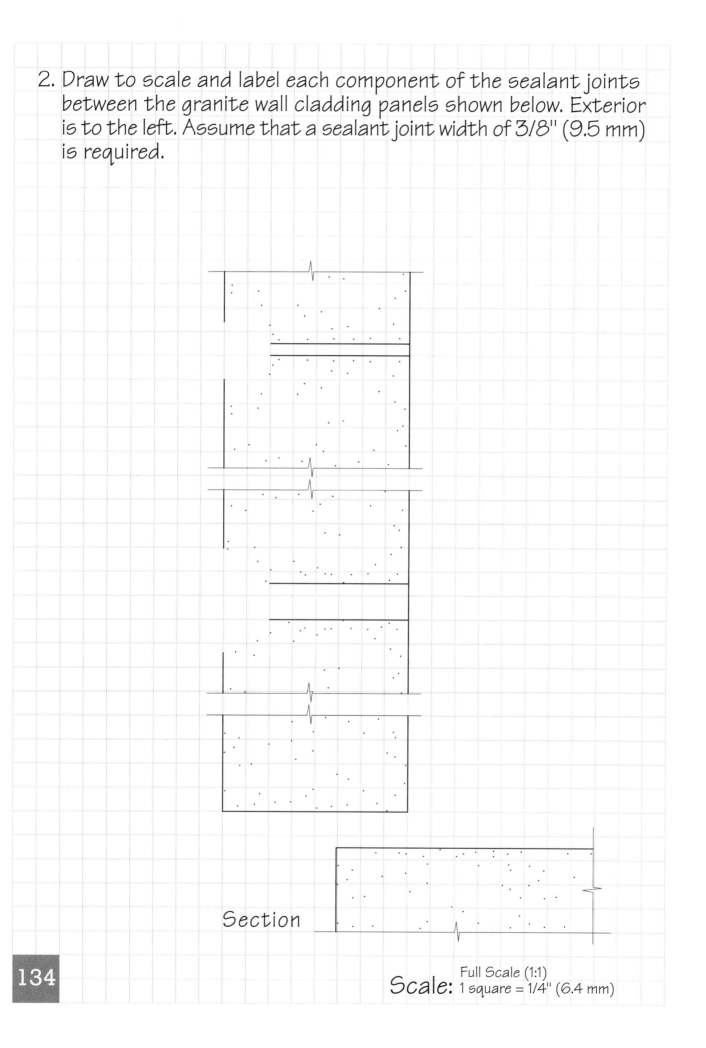

Section

Scale: Full Scale (1:1)
1 square = 1/4" (6.4 mm)

20

CLADDING WITH MASONRY AND CONCRETE

20.1 Masonry Cladding
     Design

In preparation for this exercise, review Chapters 19 and 20 of the text, paying particular attention to pages 734 through 741 and the related figures.

1. To the exterior wall section below, add and label all components needed to make a watertight wall, with 1" (25 mm) of insulation in the cavity.

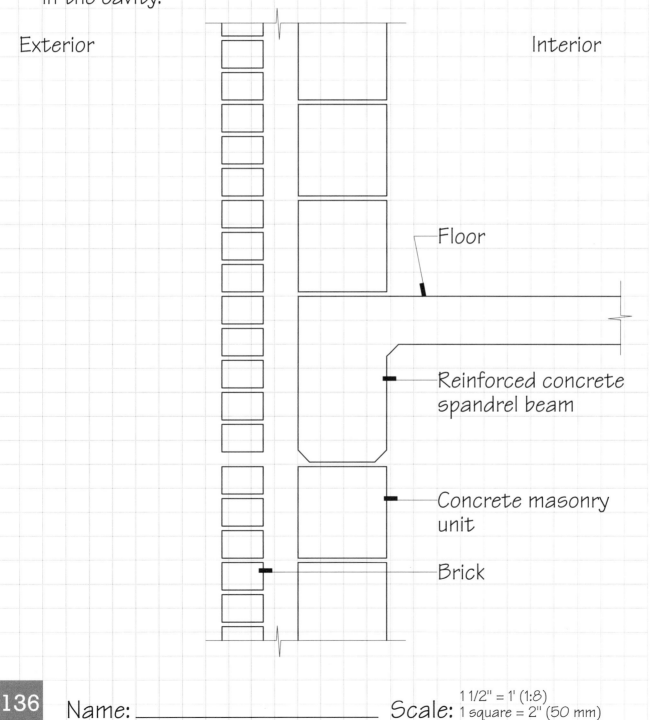

Exterior

Interior

Floor

Reinforced concrete spandrel beam

Concrete masonry unit

Brick

Name: _____

Scale: 1 1/2" = 1' (1:8)
1 square = 2" (50 mm)

CLADDING WITH METAL AND GLASS

21.1 Aluminum
     Extrusions

Aluminum Extrusions

In preparation for this exercise, review Chapter 21 of the text, to gain a good working knowledge of metal and glass curtain walls. For problems requiring design of aluminum extrusions, see especially pages 764 through 765.

1. This is a full-scale section of an extruded aluminum curtain wall vertical mullion.

   a. Name and explain the function of each of the lettered parts.

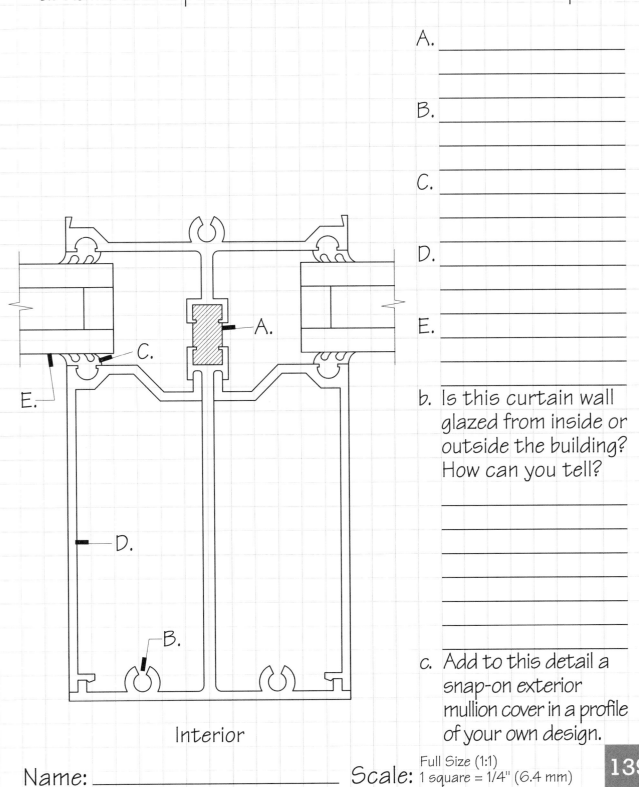

Interior

A. _____
_____
_____

B. _____
_____
_____

C. _____
_____
_____

D. _____
_____
_____

E. _____
_____

b. Is this curtain wall glazed from inside or outside the building? How can you tell?

_____
_____
_____
_____
_____
_____

c. Add to this detail a snap-on exterior mullion cover in a profile of your own design.

Name: _____

Scale: Full Size (1:1)
1 square = 1/4" (6.4 mm)

139

2. You have been asked to design an extruded aluminum post system to support 3/4" (19 mm) plywood panels on which an exhibition of photographs will be mounted. Each post should be 3" (76 mm) square and able to support one or two panels on each of its faces. A full-scale plan view of a post has been started below, with the positions of the edges of three panels shown. Complete the detail of the support system by designing and drawing features that allow the panels to be screwed or clamped to the post. No gaskets or sealants are required--the plywood may contact the aluminum directly.

Scale: Full Size (1:1)
1 square = 1/4" (6.4 mm)

SELECTING INTERIOR FINISHES

22.1 Selecting Interior Finish Systems

Parts 1 and 2 of this exercise will give you practice applying building code provisions to the selection of interior finish materials. To answer the questions, you will need to refer to Figures 1.2, 18.28, 22.5 and 22.6 of the text.

For the purposes of this assignment, assume the following fire-resistance rating requirements for various nonloadbearing elements.

Building Element	Fire-Resistance Rating
Shaft Enclosures	Within individual residences: none Connecting two or three stories: 1-hour Connecting four or more stories: 2-hour
Exit Stair Enclosures	Same as shafts
Corridors	Without sprinkler system: 1-hour With sprinkler system: none
Partitions	Separating dwelling units: 1-hour Separating mall tenants: 1-hour Separating guest rooms: 1-hour Others: none
Nonbearing, General	Where the Construction Type is noncombustible, all partitions, suspended ceilings, and other nonloadbearing elements must be noncombustible as well.

In **Part 3**, you will propose interior finish materials that can contribute toward a US Green Building Council LEED rating. Information on sustainable materials is included toward the end of many chapters in the text, and a copy of the LEED Rating System checklist is provided in Figure 1.7. More in-depth information on the LEED Rating System and how credits are applied can be fround on the U.S. Green Building Council web site, located at http://www.usgbc.org/ at the time of this writing.

Unless otherwise instructed by your teacher, assume the following for this assignment.

Building height: 5 stories   Fire protection: Unsprinklered
Occupancy Group: B   Construction Type: III-A

1. What is the Class and maximum flame spread rating permitted for each of the following?

    a. Finishes within the rooms of the building   _____
    b. Finishes in exit stairways   _____
    c. Finishes in corridors providing access to these
       stairways   _____

2. What are the required fire resistance ratings for the following parts of this building? Give units as well as numbers:

    a. Exterior loadbearing walls   _____
    b. Interior loadbearing walls   _____
    c. Columns   _____
    d. Floor construction   _____
    e. Roof construction   _____
    f. Elevator shaft enclosures   _____
    g. Elevator doors   _____
    h. Exit stair enclosures   _____
    i. Doors opening into exit stairs   _____
    j. Corridor walls   _____
    k. Doors opening into corridors   _____
    l. Nonbearing interior partitions   _____
    m. A wall separating the given occupancy from a
       Group M store occupancy in the same building   _____
    n. A door in the occupancy separation above   _____

Name: _____

3. In the space below, prepare a list of interior finish materials suitable for an office building that can help earn points toward US Green Building Council LEED certification. Also indicate the specific LEED Checklist credit(s) satisfied by each material. If you have access to product information in your school library or on the Web, include an example of a commercially available product as well. One example has been completed for you.

Material	LEED Credit & Example Product
Low-VOC carpet glue	Indoor Environmental Quality Credit 4.1 Low-Emitting Materials, Adhesives & Sealants Product: Re-Source Premium Multipurpose Adhesive 1000 (low-odor, zero-VOC)

INTERIOR WALLS AND PARTITIONS

23.1 Detailing Wall
Finishes

Detailing Wall Finishes

In preparation for doing Exercise 23.1, review Chapter 23 of the text, paying particular attention to Figures 23.13, 23.14, 23.23, 23.31, 23.32, and 23.36.

1. The plan view below shows partition framing made of fire-retardant treated 2x4 (38 mm x 89 mm) wood studs. Add 3-coat plaster surfaces over expanded metal lath to both sides. Show and label the lath, all coats of plaster, and trim accessories. Add wood finish casings to the door frame.

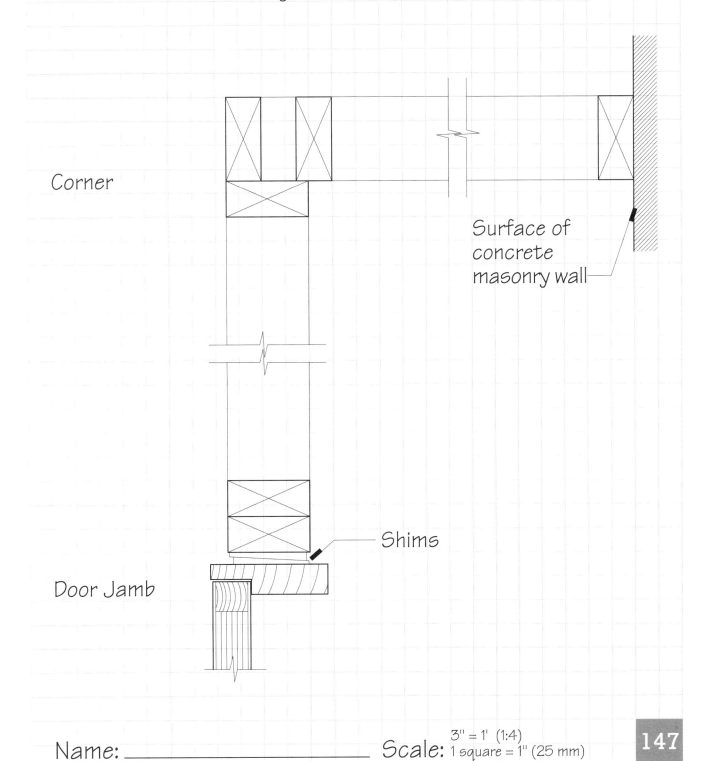

Corner

Surface of concrete masonry wall

Shims

Door Jamb

Name: _____   Scale: 3" = 1' (1:4)
1 square = 1" (25 mm)

2. The plan view on this page shows partition framing of light gauge steel studs. Add 1/2" (13 mm) gypsum board surfaces to both sides of the studs, showing and labeling all fasteners and accessories. Note that the corner is framed in the manner recommended by gypsum board manufacturers, which leaves the inside edges of the gypsum board joining at a line about 2" (51 mm) from the last stud on each wall.

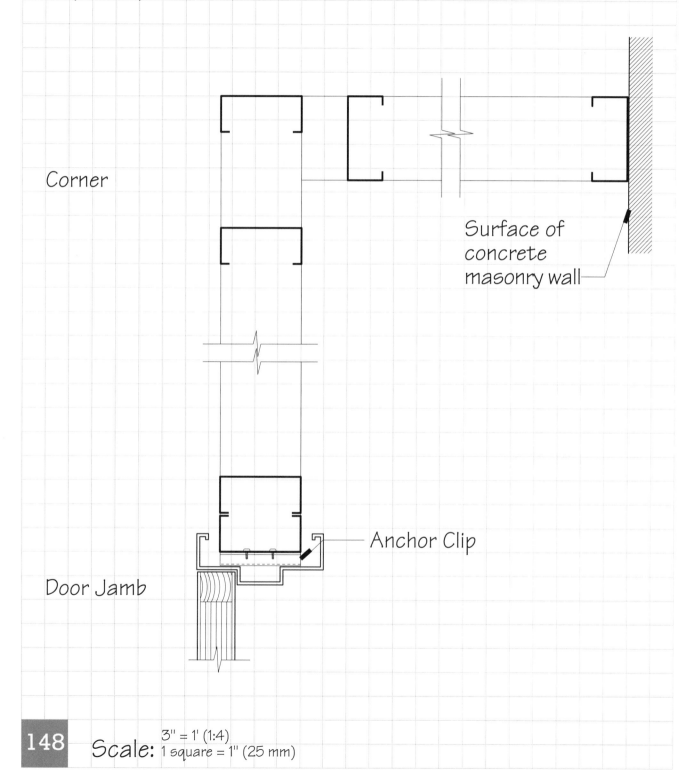

Corner

Surface of concrete masonry wall

Anchor Clip

Door Jamb

Scale: 3" = 1' (1:4)
1 square = 1" (25 mm)

FINISH CEILINGS AND FLOORS

24.1 Detailing Floor and Ceiling Finishes

Detailing Floor and Ceiling Finishes

Figures 24.3, 24.4, 24.32, and 24.35 in the text will be particularly helpful in completing this exercise.

1. On the left-hand face of this section view of a 2x4 (38 mm x 89 mm) wood stud partition, draw a veneer plaster finish. On the right-hand face, draw plaster on gypsum lath. Then add a suspended plaster ceiling on metal lath at the level indicated, and a hardwood floor on sleepers with a wood base. Show and label all coats of plaster, trim accessories, gypsum board, and other components.

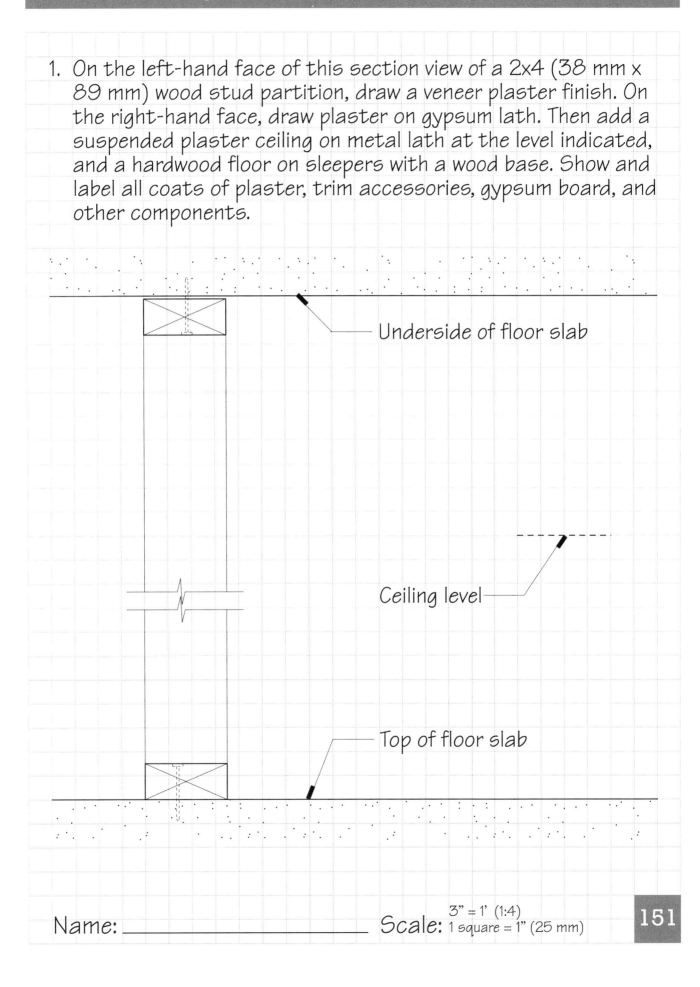

Underside of floor slab

Ceiling level

Top of floor slab

Name: _____

Scale: 3" = 1' (1:4)
1 square = 1" (25 mm)

151

2. Add 1/2" (13 mm) gypsum board surfaces to both faces of this light-gauge steel stud partition. Then draw a suspended acoustical tile ceiling with an exposed grid. Assume that the tiles are one inch (25.4 mm) thick and the grid tees are two inches (51 mm) deep. Finally, install a vinyl-composition tile floor with vinyl cove base. Show and label all trim accessories and other components.

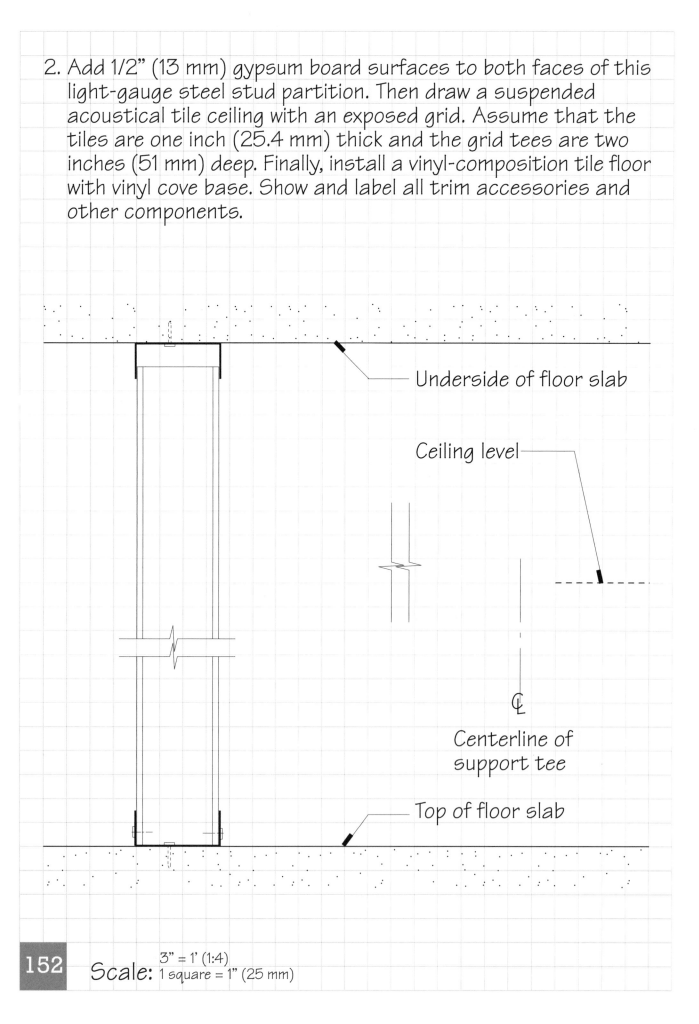

Underside of floor slab

Ceiling level

Centerline of support tee

Top of floor slab

Scale: 3" = 1' (1:4)
1 square = 1" (25 mm)

These exercises can only introduce you to the pleasures, potentials, and challenges of building construction. But your further education in construction lies all around you, ready for the taking. Here are some suggestions.

1. Never pass by a building under construction without noting carefully what materials are being used, how they are being put together, and what result is being achieved. Completed buildings, regardless of age, are also valuable as sources of information on materials and techniques, if you develop the habit of looking closely. What do you like about a given building, and how was this result obtained? Where has the building failed (a leak, a sag, a crack, an unpleasant room, an ugly exterior) and why?

2. Skilled tradespeople are the finest source of information on their particular crafts. Watch how they work, and ask questions whenever you can. In most cases a skilled worker is flattered that someone will take an interest in his/her artistry, and will be happy to talk. Even when you are the designer of a building that is under construction, listen carefully to what the workers have to tell you. Seven times out of ten they'll teach you something and your next building will be better for it.

3. Never spend only the time in a hardware or building supply store that it takes to make your purchase. Browse, and marvel at the human ingenuity that is distilled in the tools and building components you find there. Lumberyards, brickyards, quarries, fabrication shops, even gravel pits are goldmines of information on building. Use all your senses to gather this information--touch, smell, sound, sight. Become familiar with colors, odors, densities, textures, patterns, and sounds of various materials. Develop a tactile "feel" that becomes a natural part of your design know-how.

4. Read manufacturers' catalogs and literature. Send off to companies whose ads you see in architectural and engineering magazines for their literature, and start your own files of information. Visit websites of building material manufacturers and contractors. Interrogate salespersons and representatives of building materials manufacturers and suppliers whenever you meet up with them. Learn to discriminate the genuine, durable, attractive products from the shabby imitations.

5. Look for summer and part-time jobs in construction, or in the offices of architects and engineers. Pester your employers to let you work in all facets of the job, both in the office and in the field.

6. Best of all, build with your own hands, even if it is just to patch cracked plaster or fix a wobbly chair. A garage, deck, or house addition is worth an advanced degree. Read the how-to books, do the design, order the materials, and do the work. Buy good tools (a solid, lifetime investment) and keep them sharp and clean. Feel the satisfaction of each day's accomplishment. Learn from your mistakes as well as your successes. Do better next time. Yes, there will be a next time. Construction is habit forming.